Pocket Guide to
Technical Writing

P9-DXL-706

Pocket Guide to
Technical Writing

Second Edition

William S. Pfeiffer
Southern Polytechnic State University

Prentice
Hall

Upper Saddle River, New Jersey
Columbus, Ohio

Library of Congress Cataloging-in-Publication Data

Pfeiffer, William S.
  Pocket guide to technical writing / William S. Pfeiffer.—2nd. ed.
    p. cm.
  Includes index.
  ISBN 0-13-026102-5 (pbk.)
  1. English language—Technical English—Handbooks, manuals, etc. 2. English
language—Rhetoric—Handbooks, manuals, etc. 3. Technical writing—Handbooks,
manuals, etc. I. Title.

PE1475 .P465 2001
808'.0666—dc21   00-027689

Vice President and Publisher: Dave Garza
Editor in Chief: Stephen Helba
Acquisitions Editor: Debbie Yarnell
Developmental Editor: Ohlinger Publishing Services
Production Editor: Louise N. Sette
Production Supervision: Carlisle Publishers Services
Design Coordinator: Robin G. Chukes
Cover Designer: Becky Kulka
Cover photo: ©SuperStock
Production Manager: Brian Fox
Marketing Manager: Chris Bracken

This book was set in Meridien by Carlisle Communications, Ltd. It was printed and bound by
R.R. Donnelley & Sons Company. The cover was printed by Phoenix Color Corp.

Prentice
Hall

10 9 8 7 6 5 4 3 2 1
ISBN 0-13-026102-5

This book first is dedicated to the memory of my parents, Hal and Jean Pfeiffer, who wrote with honesty, brevity, and style. It is also dedicated to my children, Zachary and Katie, who are learning to do the same.

Brief Contents

Contents

CHAPTER **2** **Structure:**
Achieving Order and Design   **19**

CHAPTER **3** **Special Topics:**
Graphics and Speeches   **35**

Speeches   57

Guidelines for Preparation and Delivery   52
Guidelines for Speech Graphics   59
Guidelines for Overcoming Nervousness   62

Appendix A:
ABC Formats and Examples   65

Preface

*T*his little book aims to provide you with a handy reference for on-the-job writing. It does not pretend to be the last word on the subject. There are plenty of longer books to serve that purpose. Instead, it gives you quick, easy-to-read answers to common writing problems you will face in school and on the job. This introduction includes information on the organization of the text, changes in this edition, and the sources used in writing it.

ORGANIZATION OF TEXT

The *Pocket Guide* presents information on five main subjects, each with its own chapter or appendix:

Chapter 1   The Writing Process: Achieving Speed and Quality

Chapter 2   Structure: Achieving Order and Design

Chapter 3   Special Topics: Graphics and Speeches

Appendix A: ABC Formats and Examples

Appendix B: Writing Handbook

Each section gives you a "jump start" on planning, drafting, or revising when you are rushed to deliver a document. The two appendices will be especially useful for immediate help with writing. Appendix A includes one-page outlines ("ABC Formats") for 16 different types of documents, followed by model examples. Appendix B contains alphabetized entries to assist with questions about style, grammar, and usage.

CHANGES IN THE SECOND EDITION

New to this edition is the third chapter, "Special Topics," which includes guidelines for graphics and speeches.

Further, there are now two appendices in the book, instead of the one that appeared in the first edition. Appendix A: ABC Formats and Examples has been expanded to include job letters and resumes. Appendix B: Writing Handbook now contains an entry on English as a Second Language (ESL). The many international students in technical writing classrooms suggest that this new entry will be useful.

ACKNOWLEDGMENTS

I would like to thank the following individuals and companies for allowing me to incorporate their material into examples: Jefferson Alexander, Natalie Birnbaum, David Cox, Jeffrey Daxon, Rob Duggan, Fugro-McClelland, Inc., Chuck Keller, Becky Kelly, Steven Knapp, Law Companies Group, Jeff Orr, Chris Owen, Ken Rainey, Sam Randell, Herb Smith, and James Stephens.

Additionally, I would like to thank the reviewers of this text for their attention and insight: Genie Babb, University of Alaska, Anchorage; Debra C. Boyd, Winthrop University; and Megan O'Neill, Stetson University.

Also, many thanks to my family—Evelyn, Zach, and Katie—for their ever-present support during this project and for their outstanding suggestions for improving the book.

REQUEST FOR YOUR COMMENTS

In using the *Pocket Guide,* you may come up with suggested changes for another edition. Please send your ideas to me at the Humanities and Technical Communication Department, Southern Polytechnic State University, 1100 S. Marietta Pkwy., Marietta, GA 30060. I'd be glad to hear from you.

ABOUT THE AUTHOR

Sandy Pfeiffer has spent a career in technical and professional writing. With a Ph.D. in English, he is a professor and department head at Southern Polytechnic State University in Marietta, Georgia. His department offers both B.S. and M.S. degrees in Technical and Professional Communication.

Besides the *Pocket Guide,* Pfeiffer has written *Technical Writing: A Practical Approach,* now in its 4th edition (Prentice Hall, 2000). The book is used in colleges and universities around the country. He also wrote *Proposal Writing: The Art of Friendly Persuasion* (Merrill, 1989), and co-authored, with Chuck Keller, *Proposal Writing: The Act of Friendly and Winning Persuasion* (Prentice Hall, 2000).

Since 1979, Pfeiffer has taught communication seminars to business, industry, and government groups in the United States and overseas. Sample seminar topics include Report Writing, Proposals, Oral Presentations, Effective Meetings, and Technical Editing.

You can contact Sandy Pfeiffer at the following address:

William S. Pfeiffer, Ph.D., Professor and Head
Humanities and Technical Communication Department
Southern Polytechnic State University
1100 S. Marietta Pkwy.
Marietta, GA 30060
770-528-7206
email address: pfeiffer@spsu.edu

1 The Writing Process: Achieving Speed and Quality

*A*ll writers want to write quickly and well. If there were magic pills to create good writing on demand, they would outsell aspirin. Though we have no prose-producing pills, we do have simple techniques for dramatically increasing the speed and quality of your on-the-job writing. This chapter presents techniques to improve the writing process.

The "writing process" can be defined as the steps you follow to complete a successful writing project. Process is indeed the key to good writing because it separates activity into manageable stages, each of which includes specific goals. To introduce you to the writing process, this chapter has three main sections:

- **Definition of Technical Writing**—giving basic information about purpose, writers, and readers in technical writing
- **Nine Steps to Better Writing**—explaining the main steps that help you plan, draft, and revise
- **Writing in Groups**—providing five guidelines for collaborative writing projects

The chapter ends with a writing checklist (Figure 1–7) for you to use while you write.

DEFINITION OF TECHNICAL WRITING

At some point in our lives, we all do three main types of writing: academic, personal, and technical (see Figure 1–1). Yet for most mortals, only technical writing remains the type that will determine our professional success.

The term *technical writing* includes all written communication done on the job. It originally referred only to writing done in fields of technology, engineering, and science, but it has come to mean writing done in *all* professions and organizations (see Figure 1–2). Technical writing can be distinguished from other prose by features related to its (1) purpose, (2) writer, and (3) readers.

TYPE	PURPOSE	AUDIENCE	EXAMPLE
Academic	Display knowledge	Teachers or colleagues	Research paper
Personal	Enlighten, entertain	Yourself or friends	Journal, letters
Technical	Get something done	Supervisors, subordinates, or customers	Reports

FIGURE 1–1
Three types of writing

■ PURPOSE: GETTING SOMETHING DONE

A practical purpose underlies all on-the-job writing. With such writing you strive to get something accomplished for your organization, for a customer, or for both. Academic writing displays knowledge, and personal writing entertains or enlightens. Although technical writing may sometimes have these goals too, its main purpose is *practical*—for example, to change a policy, offer a new product, or explain a procedure.

■ WRITER: CONVEYING YOUR KNOWLEDGE TO THE READER

As writer, you have something to teach your readers. Usually you know more about the topic than they do—that's why *you're* writing to *them*. Readers benefit from knowledge you provide and make changes accordingly—for example, responding to problems you highlight, following recommendations you put forth, and buying products you are selling. Of course, your superior knowledge about the topic can sometimes create a problem. You must avoid talking over the heads of your readers. This challenge may be the greatest one you face as a writer.

> Writer's Job #1: Write for the Reader

■ READERS: UNDERSTANDING THEIR DIVERSE NEEDS

Your job would be simple if each document were directed to just one reader. But actual technical writing is not that easy. Instead, a document often has *many* readers, with mixed technical backgrounds and with different needs. Even when a document goes to only one person, it may be read by other people later. This varied audience affects the structure and the language you select to drive home your message.

Correspondence: In-House or External
- Memos to your boss and to your subordinates
- Routine letters to customers, vendors, etc.
- "Good news" letters to customers
- "Bad news" letters to customers
- Sales letters to potential customers
- Electronic mail (email) messages to coworkers or customers over a computer network

Short Reports: In-House or External
- Analysis of a problem
- Recommendation
- Equipment evaluation
- Progress report on project or routine periodic report
- Report on the results of laboratory or field work
- Description of the results of a company trip

Long Reports: In-House or External
- Complex problem analysis, recommendation, or equipment evaluation
- Project report on field or laboratory work
- Feasibility study

Other Documents
- Proposal to boss for new product line
- Proposal to boss for change in procedures
- Proposal to customer to sell a product, service, or idea
- Proposal to funding agency for support of research project
- Abstract or summary of technical article
- Technical article or presentation
- Operation manual or other manual

FIGURE 1–2
Examples of technical writing
Source: Technical Writing: A Practical Approach, 4th ed. (p. 7) by W. S. Pfeiffer, 2000, Upper Saddle River, NJ: Prentice Hall. Reprinted by permission.

Thus the fabric of technical writing is formed by the special combination of three elements, as shown in Figure 1–3:

1. A practical purpose
2. A writer more informed than the audience
3. Readers with diverse needs

Such writing will challenge you. The rest of this chapter shows how the challenge can be met most efficiently during the planning, drafting, and revision stages of writing.

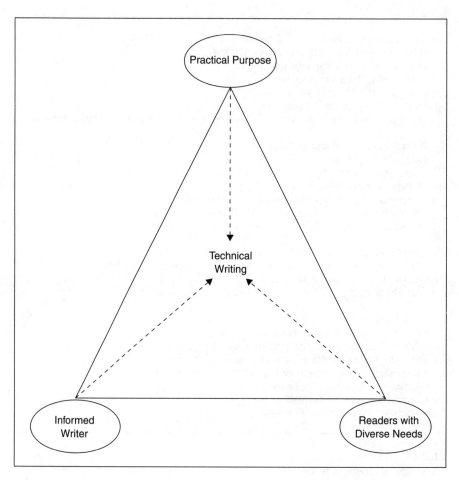

FIGURE 1–3
Technical writing triangle

NINE STEPS TO BETTER WRITING

Most good writers are made, not born. They work hard at perfecting the *process* of writing. The written *product* may appear effortlessly produced, but actually it results from a disciplined approach to composition. The discipline is embodied in a nine-step plan. If you follow this plan, you will learn for yourself that there is no mystery to improving the efficiency and quality of your writing.

Of the main stages of writing—planning, drafting, and revising—the first deserves the most attention but often receives the least. An extra hour spent planning can save two or three hours of rework during the drafting and revision stages. For that reason, the first seven steps concern the planning process. Then Steps 8 and 9 offer suggestions for drafting and revising your draft. (See Figure 1–7 for a checklist that includes all nine steps.)

■ STEP 1: WRITE A BRIEF PURPOSE STATEMENT

A one- or two-sentence purpose statement helps get you started. It becomes the lead-in to your outline and then often becomes the first words in the document itself. Writing a purpose statement forces you to decide exactly why you are writing. It directs your effort and becomes the lens through which you view the entire writing project.

A purpose statement can begin with a simple phrase like "The purpose of this report is to . . . ," or it can include more subtle phrasing. Whatever wording you choose, strive to write a passage that clearly sets forth your intentions.

Purpose Statements

Example 1: This report presents the findings of our fieldwork at Trinity Dam, along with our recommendation that the spillway be replaced.

Example 2: The purpose of this report is to compare and contrast two computer systems being considered for Exton, Inc. The report draws conclusions about the system best suited for Exton's long-term needs.

Note that a purpose statement only indicates why you are writing. It does not present summary points such as conclusions or recommendations. It gives direction, not results, and intends only to get you started in the writing process.

■ STEP 2: CONSIDER THE OBSTACLES YOUR READERS FACE

In an ideal world, readers would anxiously await your documents and give them their undivided attention. That ideal rarely exists. In fact, you're better off making the following assumptions about the audience members.

They Are Always Interrupted
Meetings, phone calls, email, lunch, and even business trips may interrupt them as they attempt to read your document.

They Are Impatient
They want you to deliver the goods quickly and clearly. While they read, they are thinking, "What's the point?" or "So what?" or "What does this have to do with me?" This impatience may mean they will not, or cannot, read the document from beginning to end.

They Lack Your Technical Knowledge

They may not understand the technical language you use. Thus they'll feel insulted—or just plain lost—if you speak in terms they cannot understand.

Most Documents Have More Than One Reader

Readers share decision making with others. Your document may be read by persons from different levels and with different needs.

Keeping these points in mind, next you need to determine the technical levels and decision-making authority of your readers.

■ STEP 3: DETERMINE TECHNICAL LEVELS OF READERS

How much do your readers know about the subject? To answer this question, you can place most readers into one of the four categories that follow.

Category 1: Managers

Managers often are removed from the hands-on details of the topic. They need brief summaries, background information, and definitions of technical terms.

Category 2: Experts

Experts have a good understanding of the technical aspects of your topic. They need supporting technical detail, helpful tables and figures, and even appendices of supporting information.

Category 3: Operators

Operators—such as field technicians, office workers, or members of a sales force—apply the ideas in your document to their jobs. They need (a) clear organization so that they can find sections relevant to them, (b) well-written procedures, and (c) clarity about how the document affects their jobs.

Category 4: General Readers

General readers, also known as "laypersons," have the least amount of information on your topic and often are outside an organization—for example, citizens reading a report on the environmental impact of a factory proposed for their community. They need (a) definitions of technical terms, (b) frequent use of graphics, and (c) clear statements about how the document will affect them.

■ STEP 4: DETERMINE DECISION-MAKING LEVELS OF READERS

Your readers also can be classified by the degree to which they make decisions based on your document. During the planning process, you can classify your readers into the following three groups.

Decision-Makers

As managers, decision-makers translate your document into action. Decisions can be made by an individual or by a committee.

Advisers

Although advisers do not make decisions themselves, they have the ear of people who do. That is, busy managers often give engineers, accountants, and others the task of analyzing technical points of a document. Final decisions may rest on recommendations that flow from this analysis.

Receivers

Some readers are not part of the decision-making process. Instead, they receive information in the document and make adjustments in their jobs accordingly.

Figure 1–4 includes technical levels on the vertical axis and decision-making levels on the horizontal axis. It shows 12 basic categories of readers. Generally, the more categories that apply to any given writing project, the more challenging is your job in meeting the needs of a varied audience.

Technical Level	Decision-Making Level		
	Decision-Makers	Advisers	Receivers
Managers			
Experts			
Operators			
General Readers			

FIGURE 1–4
Reader matrix
Source: Technical Writing: A Practical Approach, 4th ed. (p. 16) by W. S. Pfeiffer, 2000, Upper Saddle River, NJ: Prentice Hall. Reprinted by permission.

■ *Step 5: Find Out What Decision-Makers Want*

Knowing the technical and decision-making levels of your readers is a good first step. Next you need to focus on the needs of the most important readers—those who make decisions. Three suggestions follow.

Write Down What You Know about Decision-Makers
Specifically, try to get answers to these six questions:

1. What is their educational background?
2. What is their technical literacy on the topic?
3. What main question do they want answered?
4. What main action do you want them to take?
5. What style or format do they like in documents?
6. What personality traits may affect their reading?

Talk with Colleagues Who Have Written to the Same Readers
Someone else in your organization may know the needs of the decision-makers for your document. Ask around the office to see if your colleagues can help you answer the six preceding questions.

Remember That All Readers Prefer Simplicity
When you just aren't able to find out much about the decision-makers, remember one essential point about all readers: They prefer documents to be as short and simple as possible. The popular KISS principle is the best rule to follow:

> Keep It Short and Simple!

■ *Step 6: Collect and Document Information Carefully*

Whether you gather information yourself or get it from other sources, be careful during the research phase of the planning process. First, your reader may want to know exactly how you developed supporting data. Second, your professional reputation—and even your job—may be at risk if you err in the way you handle borrowed information. Some suggestions follow.

Record Notes Carefully
Remember the research skills you learned in high school or college? They stressed the importance of taking careful notes on any material taken from other sources. Your notes must do the following:

■ Distinguish your summarizing from direct quotations
■ Include exact wording of direct quotations
■ Label the exact citation of the source (title, author, page, etc.)

Carefully Transfer Information from Notes to Draft

Most errors in documentation result from sloppy work. In writing drafts from notes, circle quotation marks to make obvious what you have quoted from a source. An error at this stage could produce plagiarism in the document. Plagiarism means the parading of someone else's ideas, words, or graphics as your own—whether done intentionally or unintentionally.

Use the Right Citation System

In citing borrowed information, use a documentation system most familiar to your reader. Some common systems are those endorsed by the CBE (Council of Biology Editors), APA (American Psychological Association), and MLA (Modern Language Association). Another common resource is the *Chicago Manual of Style*. Most reference systems now rely on parenthetical citations—that is, abbreviated references in the text of the document, as follows (Pfeiffer, p. 12). Then your sources would be listed at the end of the document, with complete information on each source. For example:

> Pfeiffer, William S. *Pocket Guide to Technical Writing.* 2nd ed. Upper Saddle River, NJ: Prentice Hall, 2001.

■ *Step 7: Write an Outline*

Completing an outline is the single best way to write both quickly and well. This section answers three main questions: (1) Why is an outline so important? (2) What should it look like? and (3) What steps should you use in writing it?

First, why is an outline so important? Here are the main reasons.

- **Organization:** It forces you to grapple with matters of organization at a time when it is easy to change the structure of the document—that is, before you have committed words to draft. As you add and delete ideas on the page and shift points from main to secondary topics, you are thinking about the best way to satisfy the readers' needs.
- **Visualization:** It shows you—visibly—whether you have enough supporting information. For example, if your outline includes only one subheading for a topic, you know that you need either to do more research or to delete the topic.
- **Review:** It speeds up the review process when your documents must be approved by someone else in your organization. It is much easier to make changes at this stage than later, when you have invested time in the draft. Indeed, reviewers who "sign off" on your outline are much less likely to request major changes later.

Second, what should an outline look like? You may remember a teacher requiring you to submit a perfectly neat outline with ideal format. Such an outline is never the first version, for we just don't think that neatly. Instead,

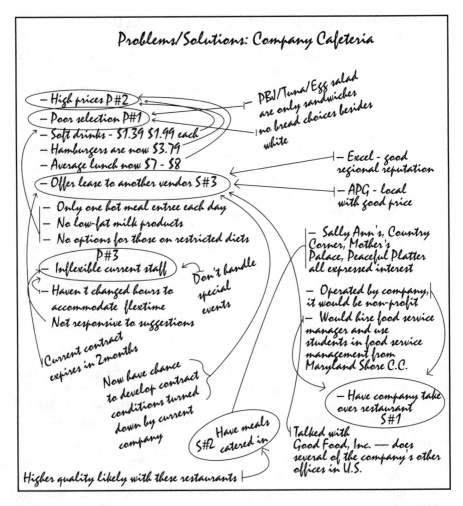

FIGURE 1–5
The outlining process: Early stage
Source: Technical Writing: A Practical Approach, 4th ed. (p. 22) by W. S. Pfeiffer, 2000, Upper Saddle River, NJ: Prentice Hall. Reprinted by permission.

each outline will (1) start as points spread over a page without any recognizable hierarchy, (2) evolve to a recognizable list of main and supporting points, and (3) change shape during the drafting process as you adjust points of emphasis. Outlining is messy business because it reflects the thinking you do *before* you write.

Third, what steps should you use in writing an outline? Follow the sequence below. Figure 1–5 shows the result of the first two steps. Figure 1–6 shows a later outline resulting from the third step.

PROBLEMS AND SOLUTIONS: CURRENT CAFETERIA IN BUILDING

I. Problem #1: Poor selection
   A. Only one hot meal entree each day
   B. Only three sandwiches—PBJ, egg salad, and tuna
   C. Only one bread—white
   D. No low-fat milk products (milk, yogurt, LF cheeses, etc.)
   E. No options for those with restricted diets
II. Problem #2: High prices
   A. Soft drinks from $1.39 to $1.99 each
   B. Hamburgers now $3.79
   C. Average lunch now $7–$8
III. Problem #3: Inflexible staff
   A. Unwilling to change hours to meet McDuff's flexible work schedule
   B. Have not acted on suggestions
   C. Not willing to cater special events in building
IV. Solution #1: End lease and make food service a McDuff department
   A. Hire food service manager
   B. Use students enrolled in food service management program at Maryland Shore Community College
   C. Operate as nonprofit operation—just cover expenses
V. Solution #2: Hire outside restaurant to cater meals in to building
   A. Higher quality likely
   B. Initial interest by four nearby restaurants
      1. Sally Ann's
      2. Country Corner
      3. Mother's Palace
      4. Peaceful Platter
VI. Solution #3: Continue leasing space but change companies
   A. Initial interest by three vendors
      1. Excel—good regional reputation for quality
      2. APG—close by and local, with best price
      3. Good Food, Inc.—used by two other McDuff offices with good results
   B. Current contract over in two months
   C. Chance to develop contract not acceptable to current company

FIGURE 1–6
The outlining process: Later stage
Source: Technical Writing: A Practical Approach, 4th ed. (p. 23) by W. S. Pfeiffer, 2000, Upper Saddle River, NJ: Prentice Hall. Reprinted by permission.

Record Random Ideas Quickly

This "nonlinear" (translate—messy) process involves the free association of ideas. To stay on track, write your purpose statement at the top of the page. Then scribble down as many points as possible that relate to the topic.

Show Relationships

Surveying your page of ideas, locate the three or four main points that indicate the direction your document will take. Circle them. Then draw lines to connect these main points to their many supporting points, as shown in Figure 1–5.

Draft a Final Outline

Your messy outline now must be cleaned up to make it useful to you as you write. You can either use the traditional format with Roman numerals, etc. (see Figure 1–6), or you can simply use dashes and indenting to indicate outline levels. Whatever format you employ, the final outline should reflect three main features:

- **Depth:** Be sure the entire outline has enough support to develop the draft.
- **Balance:** Include adequate detail for all of your main points. When you subdivide a point, have at least two breakdowns—any object divided has at least two parts.
- **Parallel Form:** Give points in the same grouping the same grammatical form, both to make the outline easier to read and to ease the transition later to parallel text in the first draft. In fact, consider keeping the entire outline consistent in form—for example, by using either full sentences or fragments for your points. Fragments are preferable because they take up less space and don't lock you into sentences too early.

With outline in hand, you are ready to begin the second part of the three-part writing process—drafting.

■ STEP 8: WRITE YOUR FIRST DRAFT QUICKLY

The first draft offers you another chance to speed up the writing process. Unfortunately, this is when many writers encounter "writer's block" and slow down. What follow are suggestions for capturing time often wasted at this stage. Of course, the most important suggestion was covered in the previous step—*always* enter the drafting stage with a complete outline in hand.

Schedule Blocks of Drafting Time

Writing requires concentration. Close your door if you have one, head to an empty office, or write at home—just do whatever you must to keep from being interrupted for at least 30 to 60 minutes. By giving writing a priority, you will finish drafts in record time. However, if you allow every other activity to upset your drafting schedule, writing even a simple document will consume your day.

Don't Stop to Edit

Editing uses a different part of your brain than drafting does. When you stop to correct grammar or spelling errors, you derail the drafting effort. With each interruption it becomes harder and harder to regain the concentration needed to write the draft. Save editing for later.

Begin with the Easiest Section

A well-structured outline gives you the luxury of starting with almost any part of the document. Skip around if doing so allows you to write the draft more quickly.

Write Summaries Last

Summary sections—like the executive summary for a formal report— usually get written last. They require the kind of thoughtful overviews and careful wording that you can write best after you've seen where the entire document is going. Of course, you can write a summary at any time if you have a good outline. It's just that you will write the *best* one when you have the chance to view a completed draft of the rest of the document.

Your goal is to get words on paper quickly. Following the suggestions above will save you hours each week, days each year, and weeks over a career. The ideas are simple, but they work.

■ *STEP 9: REVISE IN STAGES*

During revision, you must attend to matters of content, style, grammar, and mechanics. The trick to solving a variety of editing problems is to adopt a "divide-and-conquer" approach. Review the draft several times, correcting a different set of problems on each run-through. This effort to focus your attention on specific problems yields the best results. Four stages of revision follow.

Adjust and Reorganize Content

This task could be considered either the first stage of revision or the last stage of drafting. Here you expand sections that need more development, shorten some sections, and change the location of some passages.

Edit for Style

Style refers to changes that make your writing smoother, more readable, and more interesting. Here are options to consider:

- Shorten sentences
- Improve clarity—for example, by adding transitional words and by re-wording passages to show the logical flow of ideas
- Change passive-voice sentences to active
- Define technical terms
- Add headings, lists, and graphics
- Replace longer words with synonyms that are shorter or easier to understand

Edit for Grammar

Grammar haunts all of us. Your head may be filled with dozens of half-remembered rules and terms from school days—comma splices, sentence fragments, subject-verb agreement, and dangling participles, to name just a few. You want to follow the rules, but you're no English teacher and certainly don't have time to become one. The answer to this dilemma is twofold:

- First, use your own knowledge or the editorial advice of another to determine the most common grammar errors in your writing. Focus mainly on this short list—*not* on all grammar errors—when you edit.
- Second, use Appendix B of this book as an editing resource while you write.

Edit for Mechanics

Examples of mechanical errors include the following:

- Omitted words or phrases
- Misspelled words
- Inconsistent margins
- Wrong paging
- Nonparallel format in headings or subheadings

Word processing has made such errors less common and easier to correct. But technology also has lulled us into a false sense of security. You need to remain vigilant in spotting the potentially embarrassing errors that may have worked their way into your draft.

In summary, these nine steps comprise a strategy for writing every kind of on-the-job document. They require you to apply the same degree of organization to writing that you apply to other parts of your job. After all, how much of your average day do you spend writing? 20%? 30%? 50%? Using a disciplined method will help you use that time more efficiently, while producing better work.

Figure 1–7 compresses the nine guidelines in this chapter into a writing checklist to use for any document.

WRITING IN GROUPS

Group writing is common in today's organizations, which emphasize teamwork at all levels and in all activities. For example, a successful proposal or technical manual may result from the combined effort of technical specialists, marketing experts, graphic artists, writers, editors, and word processors.

Directions: The following checklist addresses the nine guidelines in this chapter. The only change here is that the guidelines are written as questions, for you to consider as you are writing.

PLANNING

1. Have you written a brief purpose statement? ______
2. Have you considered the obstacles your readers face? ______
3. Do you know the technical levels of your readers? ______
4. Do you know the decision-making levels of your readers? ______
5. Do you know exactly what decision-makers want? ______
6. Have you collected and documented information carefully? ______
7. Have you written an outline? ______

DRAFTING

8. Have you written the first draft quickly? ______

REVISING

9. Have you revised in stages? ______

FIGURE 1–7
Writing checklist

At its best, group writing benefits from the collective experience and specialties of all participants. This section describes five pointers for keeping a collaborative writing project on track:

1. Get to know your group
2. Set clear goals and ground rules
3. Use brainstorming techniques
4. Agree on a revision process
5. Use computers to communicate

Guideline 1: Get to Know Your Group

If you're used to being a "lone ranger"—that is, a solitary writer—you have some adjustments to make in reaching consensus during a group project. The best way to open the channels of communication is to learn as much as possible about your colleagues. Here are two techniques for doing so:

- Spend time talking to group members before the project begins, establishing a personal relationship, or
- Use the first session for informal chatting, before you get down to business.

The point is this: The success of your group project may depend as much on having a good working relationship as it does on the technical specialties each member brings to the table.

■ GUIDELINE 2: SET CLEAR GOALS AND GROUND RULES

All groups should set goals and establish operating rules. Specifically, the following questions must be answered either before or during the first meeting:

- What is the main objective?
- How will tasks be distributed?
- How will conflicts be resolved?
- What's the schedule for the work?

■ GUIDELINE 3: USE BRAINSTORMING TECHNIQUES

The most common error in group work is getting too critical too quickly. Just as an individual writer needs the chance to spill ideas onto paper during the outline stage, members of a group also need a chance to share ideas openly without concern for criticism. That opportunity for a non-judgmental pooling of ideas—or "brainstorming"—comes at the beginning of the group's work. Here's one strategy for brainstorming:

1. Ask the group recorder to take down ideas as quickly as possible
2. Write ideas on large pieces of paper affixed to the walls in a room
3. Use ideas written on the wall sheets to suggest additional ideas
4. Distribute results of the exercise to group members after the meeting
5. Meet again to choose ideas of most use in the project

■ GUIDELINE 4: AGREE ON A REVISION PROCESS

Group drafting and editing can be difficult. Some people used to writing in solitude find it hard to reach consensus on matters of style. Follow these suggestions for keeping the collaborative process on track:

1. Avoid making changes for the sake of one individual's preference
2. Search for areas of agreement among members, not areas of disagreement
3. Make only those style and grammar changes that can be supported by rules of style, grammar, and word use (see Appendix B)
4. Ask the group's all-around best stylist to complete a final edit

Your goal should be a document that sounds as much as possible as if it were written by one person.

■ GUIDELINE 5: USE COMPUTERS TO COMMUNICATE

Whether group members are in the same building or spread out across the world, they can benefit from using computers to communicate about the project. This technology, though changing as I write this book, falls into two general categories:

- **Asynchronous:** Such software permits members to post messages to each other and to alter a document, though not at the same time that other group members are making contributions.
- **Synchronous:** Software of this type allows group members to carry on simultaneous computer "conversations" about a document in real time. With several windows on their screens, they can place messages on the screen at the same time they edit a document that appears on the screen of all participants involved in the computer conversation.

Computer conversations can go well beyond email. Perhaps the greatest advantage they lend to group work is the openness encouraged by such on-line discussions. When coordinated with face-to-face sessions, such computer meetings produce results.

2 Structure: Achieving Order and Design

*G*ood structure will impress your readers. They respond well to inviting, easy-to-read documents. The purpose of this chapter is to help you apply simple rules about structure to everything you write.

Structure is defined here as both the arrangement of information within the document (*organization*) and the techniques used to highlight information (*page design*). This chapter includes guidelines on both aspects of structure. Three main sections follow:

1. **Basics of Organization**—describes the main rule of tech writing and the three principles that flow from it
2. **ABC Format**—outlines a three-part pattern of organization—abstract, body, conclusion—that applies to all technical documents
3. **Page Design**—explains simple formatting techniques that can improve any technical document

Together, the three sections provide an "executive summary" about structure in technical writing. Refer to Appendix A, which includes ABC Format outlines and accompanying models for 16 common documents. It can serve as a starting point for your on-the-job writing projects.

BASICS OF ORGANIZATION

As noted in Chapter 1, readers may differ greatly in technical background and decision-making authority. Yet most share four features: (1) they are interrupted while reading, (2) they are impatient to find important information, (3) they lack your technical knowledge, and (4) the documents they read are seen by others too. Any principles of organization must respond to this set of features, for there is one cardinal rule of all technical writing:

> Write for your reader, not for yourself.

This rule sounds sensible enough. However, following it will challenge you when you have many readers with varying needs. Sometimes it is hard to find the right approach. This section presents three guidelines to remember as you plan, draft, and revise documents for readers of mixed technical backgrounds.

■ STRUCTURE RULE 1: WRITE DIFFERENT PARTS FOR DIFFERENT READERS

Most of us avoid reading even short documents straight through. Instead, we scan key parts—especially near the beginning and end—and then search for information of most importance. Some readers focus on general overviews, whereas others consider technical sections most important to their jobs.

In other words, you can't count on readers to digest information in the same order you present it. Fortunately, you can use this fact to your advantage by writing different sections of documents for different readers. For example, the lead-off summary for a report may respond especially to the needs of managers, whereas technical sections in the body may be directed toward different groups of experts.

Writing different parts for different readers does present one hazard. If you go too far in tailoring sections for different readers, the document may become fragmented. To remedy this problem, use common threads of organization, theme, and style to help sections hang together as one piece of work. For example, each section of a long proposal could begin with a purpose statement, emphasize related selling points, use frequent headings, and include more active- than passive-voice sentences—even though the different sections were directed toward different readers.

■ STRUCTURE RULE 2: EMPHASIZE BEGINNINGS AND ENDINGS

Most of us read fiction and essays much differently than we read technical writing. With the former, we read patiently until plot information and new knowledge come our way. With the latter, we rush to find essential information and are impatient when we don't find it quickly.

In technical writing you must accommodate impatient readers by placing important information where they want it—at the beginning and end of the document. Figure 2–1 shows how this emphasis on beginnings and endings responds to the interest level of readers. Note that interest peaks at the beginning. Here is where the reader needs you to answer the classic "so what" question:

"So what does this document mean to me?"

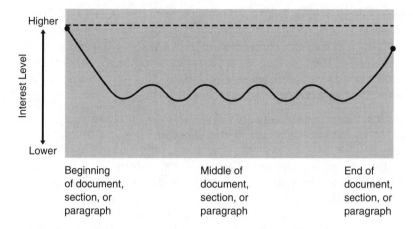

FIGURE 2–1
Reader interest curve
Source: Technical Writing: A Practical Approach, 4th ed. (p. 76) by W. S. Pfeiffer, 2000, Upper Saddle River, NJ: Prentice Hall. Reprinted by permission.

As explained in the next section, the beginning part presents just an overview of important information, not details. Specifics of major interest—like detailed lists of conclusions and recommendations—come at the end.

■ *STRUCTURE RULE 3: REPEAT KEY POINTS*

Most people will read your document selectively—not sequentially from beginning to end. This fact means that important information should be repeated, especially in longer documents.

For example, a formal report may mention a key recommendation in the cover letter, in the executive summary, and in the conclusions and recommendations section. Or a purpose statement for the report may show up, perhaps with different wording, in the cover letter, executive summary, and introduction. Needless repetition is not being advocated here. But repetition does make sense if you are delivering key information to different sets of readers.

The main principle of technical writing—write for your reader, not for yourself—and the three guidelines just covered lead directly to a pattern of organization you can use in all career writing.

ABC FORMAT

In the *Poetics,* Aristotle claimed that a literary work, to be considered "whole," must have a beginning, middle, and end. And so it is true over 2,300 years later with technical writing. Here the simple three-part structure of technical writing is labeled the ABC Format (for abstract, body, and conclusion).

Figure 2–2 gives a visual representation of the ABC Format; it also shows how the sections of the report in Figure 2–3 "map" on to this ABC Format. The diamond pattern is used because it implies the following important points about the abstract, body, and conclusion sections:

- The **Abstract** provides introductory and summary information. It is represented by the narrow top of the diamond because it is brief and leads into the Body.
- The **Body** supplies all supporting details for the document. It is represented by the broad expansive section of the diamond because it is the longest part of a document.
- The **Conclusion** gives readers what they need to act. It is represented by the narrow bottom section of the diamond because it is brief and leads away from the Body section.

In this model, note that the terms *abstract, body,* and *conclusion* refer to *generalized* parts of a document, *not* to specific headings. The exact headings used to represent these sections in documents will vary, depending on the document you're writing and the organization where you work. (See Figure 2–3.) Think of the ABC Format as a *general* organizational pattern. Within it you can fit the diverse sections of any document. What follow are guidelines for writing each of the sections. Remember that Appendix A gives ABC Formats and models for 16 common documents.

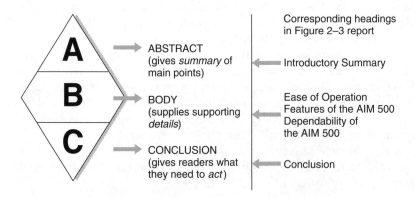

FIGURE 2–2
ABC format for all documents
Source: Technical Writing: A Practical Approach, 4th ed. (p. 79) by W. S. Pfeiffer, 2000, Upper Saddle River, NJ: Prentice Hall. Reprinted by permission.

Carruthers & Co.

MEMORANDUM

DATE: September 5, 2000
TO: Danielle Firestein
FROM: Barbara Ralston *BR*
SUBJECT: Recommendation for AIM 500 Fax

INTRODUCTORY SUMMARY

The purpose of this report is to present the results of the study you requested on the AIM 500 facsimile (fax) machine. I recommend purchase of additional AIM 500 machines, when needed, because they deliver fast, dependable service and include features we need most. This report includes the following sections: Ease of Operation, Features of the AIM 500, Dependability of the AIM 500, and Conclusion.

EASE OF OPERATION

The AIM 500 is so easy to operate that a novice can learn to transmit a document to another location in about two minutes. Here's the basic procedure:

1. Press the button marked TEL on the face of the fax machine. You then hear a dial tone.
2. Press the telephone number of the person receiving the fax on the number pad on the face of the machine.
3. Lay the document facedown on the tray at the back of the machine.

At this point, just wait for the document to be transmitted—about 18 seconds per page to transmit. The fax machine will even signal the user with a beep and a message on its LCD display when the document has been transmitted. Other more advanced operations are equally simple to use and require little training. Provided with the machine are two different charts that illustrate the machine's main functions.

The size of the AIM 500 makes it easy to set up almost anywhere in an office. The dimensions are 13 inches in width, 15 inches in length, and 9.5 inches in height. The narrow width, in particular, allows the machine to fit on most desks, file cabinets, or shelves.

FEATURES OF THE AIM 500

The AIM 500 has many features that will be beneficial to our employees. In the two years of use in our department, the following features were found to be most helpful:

Automatic redial
Last number redial memory
LCD display

FIGURE 2–3 (pp. 23–24)
Short report in ABC Format

Preset dialing
Group dialing
Use as a phone

Automatic Redial. Often when sending a fax, the sender finds the receiving line busy. The redial feature will automatically redial the busy number at 30-second intervals until the busy line is reached, saving the sender considerable time.

Last Number Redial Memory. Occasionally there may be interference on the telephone line or some other technical problem with the transmissions. The last number memory feature allows the user to press one button to automatically trigger the machine to retry the number.

LCD Display. This display feature clearly shows pertinent information, such as error messages that tell a user exactly why a transmission was not completed.

Preset Dialing. The AIM 500 can store 16 preset numbers that can be engaged with one-touch dialing. This feature makes the unit as fast and efficient as a sophisticated telephone.

Group Dialing. Upon selecting two or more of the preset telephone numbers, the user can transmit a document to all of the preset numbers at once.

Use as a Phone. The AIM 500 can also be used as a telephone, providing the user with more flexibility and convenience.

DEPENDABILITY OF THE AIM 500

Over the entire two years our department has used this machine, there have been no complaints. We always receive clear copies from the machine, and we never hear complaints about the documents we send out. This record is all the more impressive in light of the fact that we average 32 outgoing and 15 incoming transmissions a day. Obviously, we depend heavily on this machine.

So far, the only required maintenance has been to change the paper and dust the cover.

CONCLUSION

The success our department has enjoyed with the AIM 500 compels me to recommend it highly for additional future purchases. The ease of operation, many exceptional features, and record of dependability are all good reasons to buy additional units. If you have further questions about the AIM 500, please contact me at extension 3646.

FIGURE 2–3 (*continued*)

■ *ABC RULE 1: THE ABSTRACT GIVES THE "BIG PICTURE"*

The abstract component gives readers, especially decision-makers, both introductory and summary information. Specifically, it answers four questions:

1. **Purpose:** Why are you writing?
2. **Scope:** What work did you do?
3. **Results:** What main point do decision-makers want to know?
4. **Contents:** What main sections follow?

If you answer these four questions at the outset, you have given readers a capsule version of the entire document and a reason to keep reading.

In a short report or letter, the abstract might be embodied in the first few sentences. It may have a heading like "introduction" or "introductory summary" (see Figure 2–3) or it may have no heading at all. In a long report, the abstract may encompass several sections, such as the letter of transmittal, executive summary, and introduction. Remember—the term *abstract* is a construct to help you write. It can, and often does, include several document sections that, when viewed together, make up an overview.

The two paragraphs that follow satisfy the criteria for an abstract. They are from a short in-house report.

> This report describes the results of a study that compared two laser printers being considered for bulk purchase: the CopyX 3000 and Printeze 101. We studied the marketing brochures, talked to users at several firms, and tried several models loaned to us by the manufacturers.
>
> We recommend the Printeze 101. The following report supports our recommendation with information about reliability, economy of operation, and print quality. The appendix includes technical data and testimonials from three users.

■ *ABC RULE 2: THE BODY GIVES SUPPORTING DETAILS*

The document body relates details of your work. It is especially useful to technical readers who want to see support for your conclusions or recommendations. As such, it answers these kinds of questions:

- **Background:** What led up to the project?
- **Methods:** How did you gather information?
- **Data:** What information resulted from the field, lab, or office work conducted during the study?

In some documents, the body can also include detailed conclusions and recommendations—especially if such conclusions and recommendations are the main thrust of the document. In other documents, conclusions and recommendations are a brief capstone for the document and thus are reserved for a final short section. In any case, follow four general guidelines for the document body.

Use Lead-Ins at the Beginning of Sections

Lead-ins give readers a road map for what follows. They can be as simple as a list of the subsections that follow in a section. Readers need such direction finders, in the same way they need an overview at the beginning of the whole document.

Include Listings

Almost any series of three or more points should make you consider a listing. (See the "Lists" entries in Appendix B.) Bulleted or numbered lists are easier to read than long paragraphs of text, as long as you keep lists from becoming too long. Five to nine items are a maximum.

Use Graphics

Graphics draw attention to important points. They are especially useful in presenting data to technical readers, most of whom expect effective graphics.

Separate Facts from Opinions

Be clear about where opinions begin and end. Body sections usually move from facts to opinions, just like technical projects themselves. To make distinctions clear, preface your opinions with phrases like "We believe that" or "I think that."

In summary, the body sections of each document should present supporting information with clarity, with structure, and with interest.

■ *ABC RULE 3: THE CONCLUSION PROVIDES A WRAP-UP*

The conclusion component varies considerably from document to document. Generally, it answers questions like the following:

- **Results:** What are your conclusions and recommendations?
- **Action:** What happens next?
- **Emphasis:** What single point would you like to leave with readers?
- **Personal Note:** What can you add that will enhance your relationship with readers?

Because readers focus on beginnings and endings, you must exploit the opportunity provided by the conclusion part of the ABC Format.

As with the abstract and body, the conclusion is an umbrella term for any various heading labels used at the end of documents. Possible headings are Conclusions, Recommendations, Conclusions and Recommendations, Closing, and Closing Remarks. Extremely short documents, like letters, may end with a closing paragraph that has no heading. In any case, the conclusion component of the ABC Format provides the document with closure—the sense of an ending.

PAGE DESIGN

The previous section covered organization, or the arrangement of information within a document. This section deals with the other element of structure—that is, the "look" of the document, or page design.

> **Page Design:** The collection of formatting techniques used to draw attention to your writing and engage the interest of readers. Examples include use of white space, headings, lists, and varied typefaces.

Here's one way to view page design: The ABC Format forms the *deep* structure of your document, whereas page design forms the *surface* structure. Deep structure takes precedence and provides the raw material for designing pages. Yet page design can greatly influence readers. If you can capture their attention with well-designed pages, you will be more likely to engage their long-term interest in the document. This section covers three basic rules of page design.

■ DESIGN RULE 1: USE WHITE SPACE LIBERALLY

White space refers to spaces devoid of text or graphics. Empty space on the page acts like a magnet in drawing the reader's eye to text. It can also relieve visual monotony of printed words. Here are some suggestions for using white space effectively; Figure 2–4 shows these suggestions put to use.

Frame Text with 1" to 1 1/2" Margins
You may want to use an even greater margin at the bottom of the page. If your document is bound, remember to add extra margin space on the left.

Experiment with Double Columns
Long lines of text can tire the eyes, so you may want to try double columns in some documents. The extra space between the columns can help readers move down the page. However, the double-column look can also present

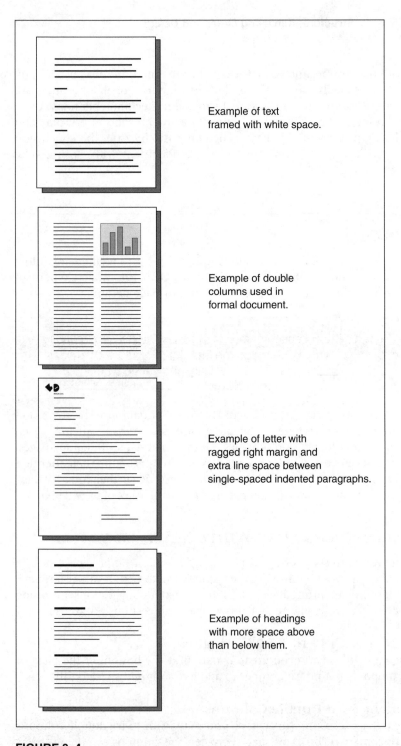

FIGURE 2–4

Use of white space

Source: Adapted from *Technical Writing: A Practical Approach*, 4th ed. (pp. 104–105) by W. S. Pfeiffer, 2000, Upper Saddle River, NJ: Prentice Hall. Reprinted by permission.

problems: (1) it may look too stilted for informal documents like short letter reports, (2) its "newspaper" appearance may encourage a cursory reading, and (3) the placement of graphics becomes challenging.

Skip Lines Between Paragraphs in Single-Spaced Text

The extra line space provides visual relief in a sea of text. You should continue to indent paragraphs, however. Most readers prefer the visual break that paragraph indenting adds.

Use Ragged Right Margins in Short Documents

The uneven edge adds visual variety needed to keep the reader's attention. Reserve a right-justified margin for some formal documents, where a book-like appearance is expected.

Use Slightly More Space Above Headings than Below Them

The additional space helps to separate a heading and related text from that which came before it.

■ *DESIGN RULE 2: USE HEADINGS AND SUBHEADINGS OFTEN*

Headings are labels used to introduce new sections and subsections. Besides helping readers stay on track, they provide visual relief from the monotony of text and assist in finding specific information later. Following are some suggestions for producing an effective heading structure.

Use Your Outline to Create Headings and Subheadings

A well-organized outline provides the headings and subheadings for the text. Use the following general rules for headings:

- Avoid single subheadings for a heading (anything divided has at least two parts)
- Maintain parallel grammatical form in headings of equal importance
- Try to have at least one heading on each page of text

Use Substantive Wording

Readers prefer headings that reflect content. Use concrete language rather than abstract nouns, even if the heading ends up being longer. Thus "Surveying the Graduates" would be preferable to "Methodology," and "Background on the Bentley Dam Failure" would be preferable to "Background."

Establish a Clear Visual Ranking of Headings

Readers should be able to identify heading levels easily as they scan a document. That is, each level should be distinctive. Figure 2–5 shows a few

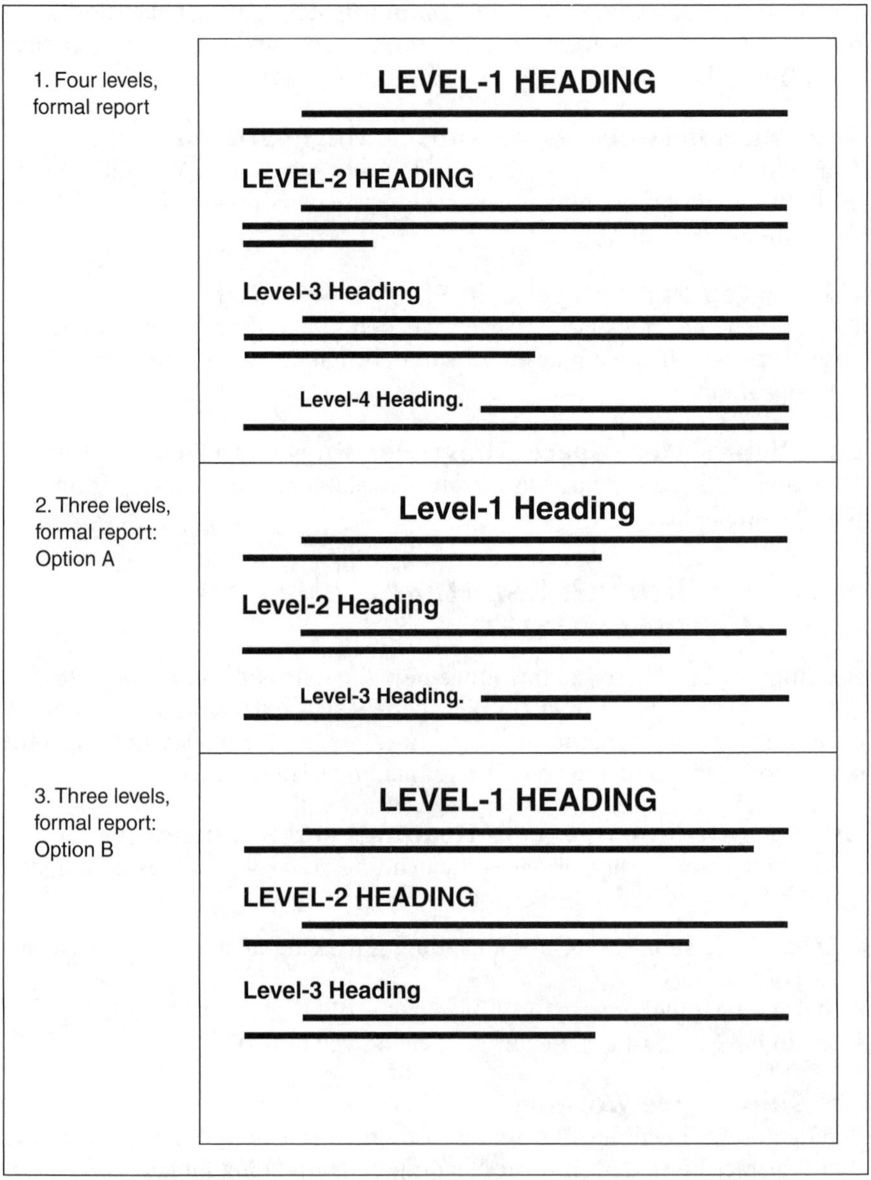

FIGURE 2–5 (pp. 30–31)
Some heading options

Source: Adapted from *Technical Writing: A Practical Approach,* 4th ed. (p. 109–111) by W. S. Pfeiffer, 2000, Upper Saddle River, NJ: Prentice Hall. Reprinted by permission.

4. Three levels, informal (letter or memo) report	**Level-1 Heading**
	━━━━━━━━━━━━━━━━━━━━━━━━━━━ ━━━━━━━━━━━━━━━━
	Level-2 Heading
	━━━━━━━━━━━━━━━━━━━━━━━━━━━ ━━━━━━━━━━━━━━
	Level-3 Heading. ━━━━━━━━━━━━━━━ ━━━━━━━━━━━━━━━━━━━━━━━━━━━
5. Two levels, informal (letter or memo) report: Option A	**LEVEL-1 HEADING**
	━━━━━━━━━━━━━━━━━━━━━━━━━━━ ━━━━━━━━━━━━━━━━━━━━
	Level-2 Heading
	━━━━━━━━━━━━━━━━━━━━━━━━━━━ ━━━━━━━━━━━━━━━━━━━━
6. Two levels, informal (letter or memo) report: Option B	**Level-1 Heading**
	━━━━━━━━━━━━━━━━━━━━━━━━━━━ ━━━━━━━━━━━━━━━
	Level-2 Heading. ━━━━━━━━━━━━━ ━━━━━━━━━━━━━━━━━━━━━━━━
7. One level, informal (letter or memo) report: Option A	**LEVEL-1 HEADING**
	━━━━━━━━━━━━━━━━━━━━━━━━━━━ ━━━━━━━━━━━━━━━━
8. One level, informal (letter or memo) report: Option B	**Level-1 Heading**
	━━━━━━━━━━━━━━━━━━━━━━━━━━━ ━━━━━━━━━━━

alternative systems for headings. There are no hard-and-fast rules for heading format. In general, keep the structure simple and neat—with a hierarchy that is clear to readers.

■ DESIGN RULE 3: USE LISTS FREQUENTLY

As a design element, lists distinguish tech writing from other types of documents like magazine essays and fiction. Almost by definition, tech writing invites you to cluster information in lists of bulleted and numbered items. Follow the suggestions below for the most effective use of lists.

Keep Lists Short
Experts say that the maximum length of a list should reflect the number of items people retain in their short-term memory—no more than five to nine items. (Three is the usual minimum for a list.) If you have more than nine items, create a grouped list with several major categories—almost in the form of an outline. This format gives readers a way to grasp information. For example, the 12 suggestions for page design in this section are grouped under three main rules.

Use Bullets and Numbers
Bullets, or enlarged dots, work best in short lists where there is no sequence. Numbers are useful when items reflect a sequence (steps in a process) or ranking (priorities for a selection). Avoid using numbers simply because a list is long, for readers will infer sequence even if you don't intend it.

Punctuate, Space, and Capitalize Lists Properly
Appendix B (Lists: Punctuation) presents several punctuation options for listings. The most common approach is (1) to avoid punctuation between items in a list made up of sentence fragments and (2) to capitalize the first word of each item in a list.

Use Proper Lead-Ins and Parallel Structure
Like headings, lists are preceded by lead-ins that should be complete grammatical units followed by colons. For example, "We have planned to" should be replaced by "We have planned the following activities:" Also, items in the same list should have parallel grammatical form. (See the Lists: General Pointers entry in Appendix B for more on parallelism.)

White space, headings, and lists are the "big three" items in page design. But there are other issues. For example, much is being written today about typefaces. You can choose from serif typefaces (with characters that have "tails" at the end of letters) or sans serif typefaces (with letters made up of lines without "tails"). The tendency is to use (1) serif type for text, where visual variety helps keep reader attention, and (2) sans serif type for headings, where the clean, uncluttered look attracts attention. In any one document, however, you should avoid the fragmenting effect of more than two or three typefaces. Beyond these basic guidelines, typeface selection remains an issue largely of individual preference.

3 Special Topics: Graphics and Speeches

*T*his chapter covers two topics of special importance to you in the workplace: graphics and speeches. Most of your documents will include graphics, so you need to know general guidelines for their use and specific rules for some common graphic types. As well, you often will be called upon to deliver oral presentations that explain facts or opinions to your managers or customers.

GRAPHICS

You do not have to be an expert to understand and use the fundamentals of graphics. In fact, the availability of high-tech graphics has made it even more important that technical professionals first understand the basics of graphics before applying sophisticated techniques. To emphasize these basics, this section (1) defines some common graphics terms, (2) explains the main reasons to use graphics and gives some general guidelines, and (3) lists specific guidelines for four common graphics.

■ *TERMS IN GRAPHICS*

Terminology for graphics is not uniform in the professions. That fact can lead to some confusion. For the purposes of this chapter, however, some common definitions are adopted and listed here:

- **Graphics:** This generic term refers to any nontextual portion of documents or oral presentations. It can be used in two ways: (1) to designate the field (for example, "Graphics is an area in which he showed great interest") or (2) to name individual graphical items ("She placed three graphics in her report").
- **Illustrations, visual aids:** Used synonymously with "graphics," these terms also can refer to all nontextual parts of a document. The term *visual aids,* however, often is limited to the context of oral presentations.

- **Tables and figures:** These terms name the two subsets of graphics.
  Tables refers to illustrations that place numbers or words in columns or rows or both.
  Figures refers to all graphics other than tables. Examples include charts (pie, bar, line, flow, and organization), engineering drawings, maps, and photographs.
- **Charts, graphs:** A subset of "figures," these synonymous terms refer to a type of graphic that displays data in visual form—as with bars, pie shapes, or lines on graphs. *Chart* is the term used most often in this text.

Of course, you may see other graphics terms. For example, some technical companies use the word *plates* for figures. Be sure to know the terms your readers understand and the types of graphics they use.

■ *REASONS FOR USING GRAPHICS*

Before deciding whether to use a pie chart, table, or any other graphic, you need to know what graphics do for your writing. Here are four main reasons for using them.

Reason 1: Graphics Simplify Ideas
Readers usually know less about the subject than you. Graphics can help them cut through technical details and grasp basic ideas. For example, a simple illustration of a laboratory instrument, such as a Bunsen burner, makes the description of a lab procedure much easier to understand. In a more complex example, Figure 3–1 uses a group of four different charts to convey the one main point—that a company's new Equipment Development group lags behind the company's other profit centers. A quick look at the charts tells the story of the group's difficulties much better than would several hundred words of text.

Reason 2: Graphics Reinforce Ideas
When a point really needs emphasis, create a graphic. For example, you might draw a map to show where computer terminals will be located within a building, or use a pie chart to show how a budget will be spent, or include a drawing that indicates how to operate a VCR. In all three cases, the graphic would reinforce points made in the accompanying text.

Reason 3: Graphics Create Interest
Graphics are "grabbers." They can be used to entice readers into the text, just as they engage readers' interest in magazines and journals. If your customers have three reports on their desks and must quickly decide which one to read first, they probably will pick up the one with an engaging picture or chart on the cover page. It may be something as simple as (1) a map outline of the state,

Problems in Equipment Development Group

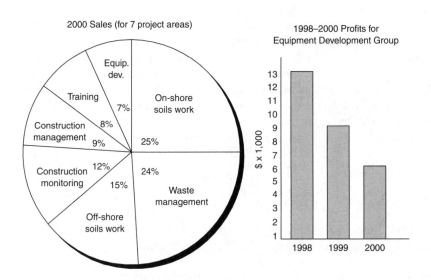

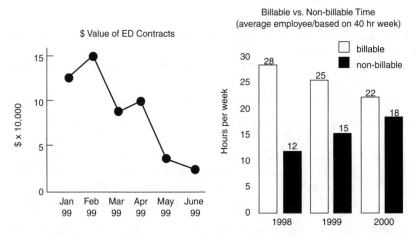

FIGURE 3–1
Graphics used to simplify ideas

county, or city where you will be doing a project, (2) a picture of the product or service you are providing, or (3) a symbol of the purpose of your writing project. Whether on the cover or in the text, graphics attract attention.

Reason 4: Graphics Are Universal

Some people wrongly associate the growing importance of graphics with to-day's reliance on television and other popular media—as if graphics pander to less-intellectual instincts. While visual media such as television obviously rely on pictures, the fact is that graphics have been mankind's universal language since cave drawings. A picture, drawing, or chart makes an immediate emotional impact that can help or hurt your case. Advertisers know the power of images, but writers of technical documents must also merge the force of graphics with their text.

■ GENERAL GUIDELINES

A few basic guidelines apply to all graphics. Keep these fundamentals in mind as you move from one type of illustration to another.

Graphics Guideline 1: Refer to All Graphics in the Text

With a few exceptions—such as cover illustrations used to grab attention—graphics should be accompanied by clear references within your text. Specifically, you should follow these rules:

- Include the graphic number in Arabic, not Roman, when you are using more than one graphic.
- Include the title, and sometimes the page number, if either is needed for clarity or emphasis.
- Incorporate the reference smoothly into text wording.

Here are two ways to phrase and position a graphics reference. In Example 1, there is the additional emphasis of the graphics title, whereas in Example 2, the title is left out. Also, note that you can draw more attention to the graphic by placing the reference at the start of the sentence in a separate clause. Or you can relegate the reference to a parenthetical expression at the end or middle of the passage. Choose the option that best suits your purposes.

- **Example 1:** In the past five years, 56 businesses in the county have started in-house recycling programs. The result has been a dramatic shift in the amount of property the county has bought for new waste sites, as shown in Figure 5 ("Landfill Purchases, 1994–1999").
- **Example 2:** As shown in Figure 5, the county has purchased much less land for landfills during the last five years. This dramatic reduction results from the fact that 56 businesses have started in-house recycling programs.

Graphics Guideline 2: Think About Where to Put Graphics

In most cases, locate a graphic close to the text in which it is mentioned. This immediate reinforcement of text by an illustration gives graphics their greatest strength. Variations of this option, as well as several other possibilities, are presented here:

- **Same page as text reference:** A simple visual, such as an informal table, should go on the same page as the text reference if you think it too small for a separate page.
- **Page opposite text reference:** A complex graphic, such as a long table, that accompanies a specific page of text can go on the page opposite the text—that is, on the opposite page of a two-page spread. Usually this option is exercised *only* in documents that are printed on both sides of the paper throughout.
- **Page following first text reference:** Most text graphics appear on the page after the first reference. If the graphic is referred to throughout the text, it can be repeated at later points. (Note: Readers prefer to have graphics positioned exactly where they need them, rather than their having to refer to another part of the document.)
- **Attachments or appendices:** Graphics can go at the end of the document in two cases: first, if the text contains so many references to the graphic that placement in a central location, such as an appendix, would make it more accessible; and second, if the graphic contains less important supporting material that would only interrupt the text.

Graphics Guideline 3: Position
Graphics Vertically When Possible

Readers prefer graphics they can view without having to turn the document sideways. However, if the table or figure cannot fit vertically on a standard 8 1/2″ × 11″ page, either use a foldout or place the graphic horizontally on the page. In the latter case, position the illustration so that the top is on the left margin. (In other words, the page must be turned clockwise to be viewed.)

Graphics Guideline 4: Avoid Clutter

Let simplicity be your guide. Readers go to graphics for relief from, or reinforcement of, the text. They do not want to be bombarded by visual clutter. Omit information that is not relevant to your purpose, while still making the illustration clear and self-contained. Also, use enough white space so that the readers' eyes are drawn to the graphic. The final section of this chapter discusses graphics clutter in more detail.

Graphics Guideline 5: Provide Titles, Notes, Keys, and Source Data

Graphics should be as self-contained and self-explanatory as possible. Moreover, they must note any borrowed information. Follow these basic rules for format and acknowledgement of sources:

- **Title:** Follow the graphic number with a short, precise title—either on the line below the number *or* on the same line after a colon (for example, "Figure 3: Salary Scales").
- **Tables:** The number and title go at the top. (As noted in Table Guideline 1 on page 50, one exception is informal tables. They have no table number or title.)
- **Figures:** The number and title can go either above or below the illustration. Center titles or place them flush with the left margin.
- **Notes for explanation:** When introductory information for the graphic is needed, place a note directly underneath the title *or* at the bottom of the graphic.
- **Keys or legends for simplicity:** If a graphic needs many labels, consider using a legend or key, which lists the labels and corresponding symbols on the graphic. For example, a pie chart might have the letters *A, B, C, D,* and *E* printed on the pie pieces, while a legend at the top, bottom, or side of the figure would list what the letters represent.
- **Source information at the bottom:** You have a moral, and sometimes legal, obligation to cite the person, organization, or publication from which you borrowed information for the figure. Either (1) precede the description with the word "Source" and a colon, or (2) if you borrowed just part of a graphic, introduce the citation with "Adapted from."

Besides citing the source, it is sometimes necessary to request permission to use copyrighted or proprietary information, depending on your use and the amount you are borrowing. (A prominent exception is most information provided by the federal government. Most government publications are not copyrighted.) Consult a reference librarian for details about seeking permission.

■ GUIDELINES FOR PIE CHARTS

Familiar to most readers, pie charts show relationships between the parts and the whole—when just approximate information is needed. Their simple circles with clear labels can provide comforting simplicity within even the most complicated report. Yet the simple form keeps them from being useful when you need to reveal detailed information. Here are specific guidelines for constructing pie charts.

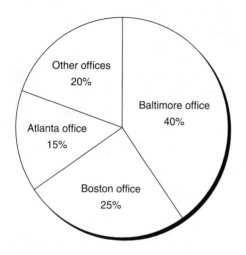

FIGURE 3–2
Pie chart with as few pieces as possible. (Chart shows work force breakdown for a project. The company can draw most project workers from its East Coast offices.)

Pie Chart Guideline 1: Use No More Than 10 Divisions

To make pie charts work well, limit the number of pie pieces to no more than 10. In fact, the fewer the better. This approach lets the reader grasp major relationships, without having to wade through the clutter of tiny divisions that are difficult to read. Figure 3–2, for example, aims to show how a large company will staff a particular project. The reader can quickly see that project staff will come from three main offices.

Pie Chart Guideline 2: Move Clockwise from 12:00, from Largest to Smallest Wedge

Readers prefer pie charts oriented like a clock—with the first wedge starting at 12:00. Move from largest to smallest wedge to provide a convenient organizing principle.

Make exceptions to this design only for good reason. In Figure 3–2, for example, the last wedge represents a greater percentage than the previous wedge. In this way, it does not break up the sequence the writer wants to establish by grouping the three offices with the three largest individual percentages of project workers.

Pie Chart Guideline 3: Use Pie Charts Especially for Percentages and Money

Pie charts catch the reader's eye best when they represent items divisible by 100, as with percentages and dollars. Figure 3–2 shows percentages; Figure 3–3 shows money. Using the pie chart for money breakdowns is made even more appropriate by the coinlike shape of the chart.

FIGURE 3–3
Pie chart showing money breakdown for average deductions
from a paycheck

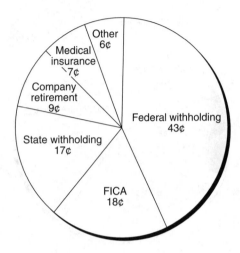

Pie Chart Guideline 4: Be Creative, but Stay Simple

Figure 3–4 shows that you can emphasize one piece of the pie by:

1. Shading a wedge
2. Removing a wedge from the main pie
3. Placing related pie charts in a three-dimensional drawing

Today there are graphics software packages that can create these and other variations for you, so experiment a bit. Of course, always make sure to keep your charts from becoming too detailed. Pie charts should stay simple.

Pie Chart Guideline 5: Draw and Label Carefully

The most common pie chart errors are (1) wedge sizes that do not correspond correctly to percentages or money amounts and (2) pie sizes that are too small to accommodate the information placed in them. Here are some suggestions for avoiding these mistakes:

- **Pie size:** Make sure the chart occupies enough of the page. On a standard 8 1/2" × 11" sheet with only one pie chart, your circle should be from 3" to 6" in diameter—large enough not to be dwarfed by labels and small enough to leave sufficient white space in the margins.
- **Labels:** Place the wedge labels either inside the pie or outside, depending on the number of wedges, the number of wedge labels, or the length of the labels. Choose the option that produces the cleanest-looking chart.
- **Conversion of percentages:** If you are drawing the pie chart by hand, not using a computer program, use a protractor or similar device. One percent of the pie equals 3.6 degrees (3.6 × 100% = 360 degrees in a circle). With that formula as your guide, you can convert percentages or cents to degrees.

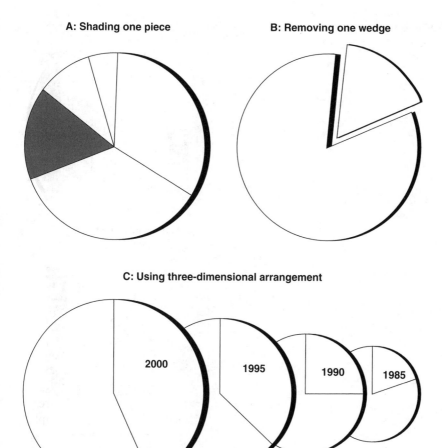

A: Shading one piece

B: Removing one wedge

C: Using three-dimensional arrangement

2000   1995   1990   1985

FIGURE 3–4
Techniques for emphasis in pie charts

Remember, however, that a pie chart does not reveal fine distinctions very well; it is best used for showing larger differences.

■ GUIDELINES FOR BAR CHARTS

Like pie charts, bar charts are easily recognized, for they are seen every day in newspapers and magazines. Unlike pie charts, however, bar charts can accommodate a good deal of technical detail. Comparisons are provided by means of two or more bars running either horizontally or vertically on the page. Follow these five guidelines to create effective bar charts.

FIGURE 3–5
Bar charts

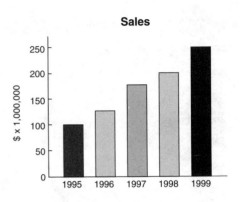

Sales

Operating Profits

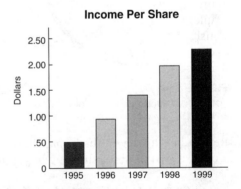

Income Per Share

Bar Chart Guideline 1: Use a Limited Number of Bars

Though bar charts can show more information than pie charts, both types of illustrations have their limits. Bar charts begin to break down when there are so many bars that information is not easily grasped. The maximum bar number can vary according to chart size, of course. Figure 3–5 shows sev-

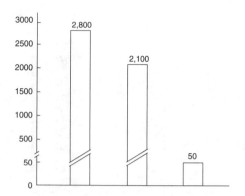

FIGURE 3–6
Hash marks on bar charts—a technique that can lead to misunderstanding

eral multibar charts. The impact of the charts is enhanced by the limited number of bars.

Bar Chart Guideline 2: Show Comparisons Clearly

Bar lengths should be varied enough to show comparisons quickly and clearly. Avoid using bars that are too close in length, for then readers must study the chart before understanding it. Such a chart lacks immediate visual impact.

Also, avoid the opposite tendency of using bar charts to show data that are much different in magnitude. To relate such differences, some writers resort to the dubious technique of inserting "break lines" (two parallel lines) on an axis to reflect breaks in scale (see Figure 3–6). Although this approach at least reminds readers of the breaks, it is still deceptive. For example, note that Figure 3–6 provides no *visual* demonstration of the relationship between 50 and 2800. The reader must think about these differences before making sense out of the chart. In other words, the use of hash marks runs counter to a main goal of graphics—creating an immediate and accurate visual impact.

Bar Chart Guideline 3: Keep Bar Widths Equal and Adjust Space Between Bars Carefully

While bar length varies, bar width must remain constant. As for distance between the bars, following are three options (along with examples in Figure 3–7):

- **Option A: Use no space** when there are close comparisons or many bars, so that differences are easier to grasp.
- **Option B: Use equal space, but less than bar width** when bar height differences are great enough to be seen in spite of the distance between bars.
- **Option C: Use variable space** when gaps between some bars are needed to reflect gaps in the data.

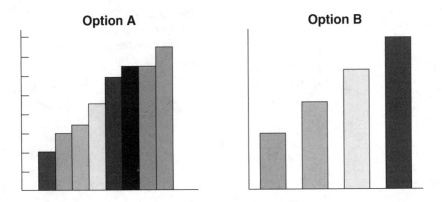

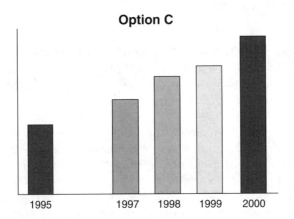

FIGURE 3–7
Bar chart variations

Bar Chart Guideline 4: Carefully Arrange the Order of Bars

The arrangement of bars is what reveals meaning to readers. Here are two common approaches:

- **Sequential:** used when the progress of the bars shows a trend—for example, a company's increasing number of environmental projects in the last five years
- **Ascending or descending order:** used when you want to make a point by the rising or falling of the bars—for example, the 1999 profits of a company's six international offices, from lowest to highest

Bar Chart Guideline 5: Be Creative

Figure 3–8 shows two bar chart variations that help display multiple trends. The *segmented bars* in Option A produce four types of information: the total

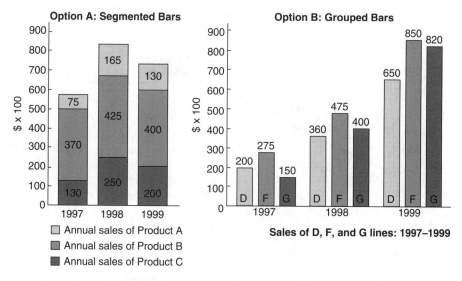

FIGURE 3–8
Bar chart variations for multiple trends

sales (A + B + C) and the individual sales for A, B, and C. The *grouped bars* in Option B show the individual sales trends for D, F, and G, along with a comparison of all three by year. Note that the amounts are written on the bars to highlight comparisons.

Although these and other bar chart variations may be useful, remember to retain the basic simplicity of the chart.

■ GUIDELINES FOR LINE CHARTS

Line charts are a common graphic. Almost every newspaper contains a few charts covering topics such as stock trends, car prices, or weather. More than other graphics, line charts telegraph complex trends immediately.

They work by using vertical and horizontal axes to reflect quantities of two different variables. The vertical (or y) axis usually plots the dependent variable; the horizontal (or x) axis usually plots the independent variable. (The dependent variable is affected by changes in the independent variable.) Lines then connect points that have been plotted on the chart. When using line charts, follow these five main guidelines:

Line Chart Guideline 1: Use Line Charts for Trends
Readers are affected by the direction and angle of the chart's line(s), so take advantage of this persuasive potential. In Figure 3–9, for example, the writer wants to show the feasibility of adopting a new medical plan. Including a line

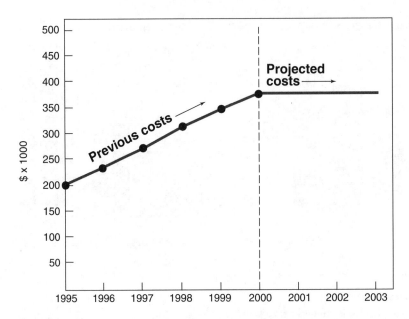

FIGURE 3–9
Line chart used to show effect of proposed medical plan

chart in the study gives immediate emphasis to the most important issue—
the effect the new plan would have on stabilizing the firm's medical costs.

Line Chart Guideline 2: Locate Line Charts with Care
Given their strong impact, line charts can be especially useful as attention-
grabbers. Consider placing them (1) on cover pages (to engage reader interest in the document), (2) at the beginning of sections that describe trends,
and (3) in conclusions (to reinforce a major point of your document).

Line Chart Guideline 3: Strive for Accuracy and Clarity
Like bar charts, line charts can be misused or just poorly constructed. Be sure
that the line or lines on the graph truly reflect the data you have used. Also,
select a scale that does not mislead readers with visual gimmicks. Here are
some specific suggestions to keep your line charts accurate and clear:

■ Start all scales from zero to eliminate the possible confusion of breaks in
  amounts (see Bar Chart Guideline 2).
■ Select a vertical-to-horizontal ratio for axis lengths that is pleasing to the
  eye (three vertical to four horizontal is common).
■ Make chart lines as thick as, or thicker than, the axis lines.
■ Use shading under the line when it will make the chart more readable.

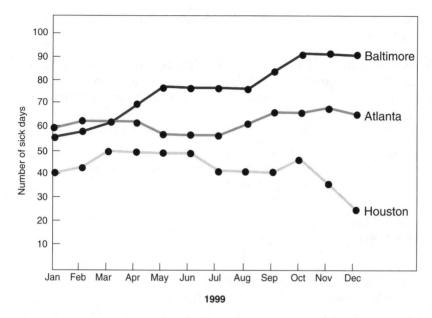

FIGURE 3–10
Line chart using multiple lines to show number of sick days taken at four offices: 1999

Line Chart Guideline 4: Do Not Place Numbers on the Chart Itself

Line charts derive their main effect from the simplicity of lines that show trends. Avoid cluttering the chart with a lot of numbers that only detract from the visual impact.

Line Chart Guideline 5: Use Multiple Lines with Care

Like bar charts, line charts can show multiple trends. Simply add another line or two. If you place too many lines on one chart, however, you run the risk of confusing the reader with too much data. Use no more than four or five lines on a single chart (see Figure 3–10).

■ GUIDELINES FOR TABLES

Tables present readers with raw data, usually in the form of numbers but sometimes in the form of words. Tables are classified as either formal or informal:

- **Informal tables:** limited data arranged in the form of either rows or columns
- **Formal tables:** data arranged in a grid, always with both horizontal rows and vertical columns

These five guidelines will help you design and position tables within the text of your documents.

Table Guideline 1: Use Informal Tables as Extensions of Text

Informal tables are usually merged with the text on a page, rather than isolated on a separate page or attachment. As such, an informal table usually has (1) no table number or title, (2) no listing in the list of illustrations in a formal report or proposal, and (3) few if any headings for rows or columns.

Example:
Our project in Alberta, Canada, will involve engineers, technicians, and salespeople from three offices, in these numbers:

San Francisco Office	45
St. Louis Office	34
London Office	6
Total	85

Table Guideline 2: Use Formal Tables for Complex Data Separated from Text

Formal tables may appear on the page of text that includes the table reference, on the page following the first text reference, or in an attachment or appendix. In any case, you should:

- Extract important data from the table and highlight them in the text
- Make every formal table as clear and visually appealing as possible

Table Guideline 3: Use Plenty of White Space

Used around and within tables, white space guides the eye through a table much better than do black lines. Avoid putting complete boxes around tables. Instead, leave one inch more of white space than you would normally leave around text.

Table Guideline 4: Follow Usual Conventions Dividing and Explaining Data

Figure 3–11 shows a typical formal table. It satisfies the overriding goal of being clear and self-contained. To achieve such an objective in your tables, follow these guidelines:

1. **Titles and numberings:** Give a title to each formal table, and place title and number above the table. Number each table if the document contains two or more tables.
2. **Headings:** Create short, clear headings for all columns and rows.
3. **Abbreviations:** Include in the headings any necessary abbreviations or symbols, such as lb or %. Spell out abbreviations and define terms in a key or footnote if any reader may need such assistance.

TABLE 6: Employee Retirement Fund			
Investment Type	Book Value	Market Value	% of Total Market Value
Temporary Securities	$434,084	434,084	5.9%
Bonds	3,679,081	3,842,056	52.4
Common Stocks	2,508,146	3,039,350	41.4
Mortgages	18,063	18,063	.3
Real Estate	1,939	1,939	nil
Totals	$6,641,313	$7,335,492	100.0%

Note: This table contrasts the book value versus the market value of the Employee Retirement Fund, as of December 31, 1999.

FIGURE 3–11
Example of formal table

4. **Numbers:** Round off numbers when possible, for ease of reading. Also, align multidigit numbers on the right edge, or at the decimal when shown.
5. **Notes:** Place any necessary explanatory headnotes either between the title and the table (if the notes are short) *or* at the bottom of the table.
6. **Footnotes:** Place any necessary footnotes below the table.
7. **Sources:** Place any necessary source references below the footnotes.
8. **Caps:** Use uppercase and lowercase letters, rather than full caps.

Table Guideline 5: Pay Special Attention to Cost Data

Most readers prefer to have complicated financial information placed in tabular form. Given the importance of such data, edit cost tables with great care. Devote extra attention to these two issues:

- Placement of decimals in costs
- Correct totals of figures

Documents like proposals can be considered contracts in some courts of law, so there is no room for error in relating costs.

SPEECHES

Oral presentations, or speeches, are defined quite broadly. Usually they can be classified according to criteria like these:

1. **Format:** from informal question/answer sessions to formal speeches

2. **Length:** from several-minute overviews to long sessions of an hour or more
3. **Number of presenters:** from solo performances to group presentations
4. **Content:** from a few highlights to detailed coverage

Throughout your career, you will speak to different-sized groups, on diverse topics, and in varied formats. The next three sections provide some common guidelines on preparation, delivery, and graphics.

■ GUIDELINES FOR PREPARATION AND DELIVERY

The goal of most oral presentations is quite simple: You must present a few basic points, in a fairly brief time, to an interested but usually impatient audience. Simplicity, brevity, and interest are the keys to success. If you deliver what *you* expect when *you* hear a speech, then you will give good presentations yourself.

Speech Guideline 1: Know Your Listeners

These features are common to most listeners:

- They cannot "rewind the tape" of your presentation, as opposed to the way they can skip back and forth through the text of a report.
- They are impatient after the first few minutes, particularly if they do not know where a speech is going.
- They will daydream and often need their attention brought back to the matter at hand (expect a 30-second attention span).
- They have heard so many disappointing presentations that they might not have high expectations for yours.

To respond to these realities, you must learn as much as possible about your listeners. For example, you can (1) consider what you already know about your audience, (2) talk with colleagues who have spoken to the same group, and (3) find out which listeners make the decisions.

Most important, make sure not to talk over anyone's head. If there are several levels of technical expertise represented by the group, find the lowest common denominator and decrease the technical level of your presentation accordingly. Remember—decision-makers are often the ones without current technical experience. They may want only highlights. Later they can review written documents for details or solicit more technical information during the question-and-answer session after you speak.

Speech Guideline 2: Use the Preacher's Maxim

The well-known preacher's maxim goes like this:

> First you tell 'em what you're gonna tell 'em, then you tell 'em, and then you tell 'em what you told 'em.

Why should most speakers follow this plan? Because it gives the speech a simple three-part structure that listeners can grasp easily. Here is how your speech should be organized (note that it corresponds to the ABC Format used throughout this text for writing):

1. Abstract (beginning of presentation): Right at the outset, you should (1) get the listeners' interest (with an anecdote, a statistic, or other technique), (2) state the exact purpose of the speech, and (3) list the main points you will cover. Do not try the patience of your audience with an extended introduction. Use no more than a minute.

Example: "Last year, Jones Engineering had 56 percent more field accidents than the year before. This morning, I'll examine a proposed safety plan that aims to solve this problem. My presentation will focus on three main benefits of the new plan: lower insurance premiums, less lost time from accidents, and better morale among the employees."

2. Body (middle of presentation): Here you discuss the points mentioned briefly in the introduction, in the same order that they were mentioned. Provide the kinds of obvious transitions that help your listeners stay on track.

Example: "The final benefit of the new safety plan will be improved morale among the field-workers at all our job sites. . . ."

3. Conclusion (end of presentation): In the conclusion, review the main ideas covered in the body of the speech and specify actions you want to occur as a result of your presentation.

Example: "Jones Engineering can benefit from this new safety plan in three main ways. . . . If Jones implements the new plan next month, I believe you will see a dramatic reduction in on-site accidents during the last half of the year."

This simple three-part plan for all presentations gives listeners the handle they need to understand your speech. First, there is a clear "road map" in the introduction so that they know what lies ahead in the rest of the speech. Second, there is an organized pattern in the body, with clear transitions between points. And third, there is a strong finish that brings the audience back full circle to the main thrust of the presentation.

Speech Guideline 3: Stick to a Few Main Points

Our short-term memory holds limited items. It follows that listeners are most attentive to speeches organized around a few major points. In fact, a good argument can be made for organizing information in groups of *threes* whenever possible. For reasons that are not totally understood, listeners seem to remember groups of three items more than they do any other size groupings—perhaps because:

- The number is simple.
- It parallels the overall three-part structure of most speeches and documents (beginning, middle, end).
- Many good speakers have used triads (Winston Churchill's "Blood, sweat, and tears," Caesar's "I came, I saw, I conquered," etc.).

Whatever the reason, groupings of three will make your speech more memorable to the audience.

Speech Guideline 4: Put Your Outline on Cards, Paper, or Overheads

The best presentations are "extemporaneous," meaning the speaker shows great familiarity with the material but uses notes for occasional reference. Avoid the extremes of (1) reading a speech verbatim, which many listeners consider the ultimate insult, or (2) memorizing a speech, which can make your presentation seem somewhat wooden and artificial.

Ironically, you appear more natural if you refer to notes during a presentation. Such extemporaneous speaking allows you to make last-minute changes in phrasing and emphasis that may improve delivery, rather than locking you into specific phrasing that is memorized or written out word for word.

Depending on your personal preference, you may choose to write speech notes on (1) index cards, (2) a sheet or two of paper, or (3) overhead transparencies. The main advantages and disadvantages of each are listed in Figure 3–12.

Speech Guideline 5: Practice, Practice, Practice

Many speakers prepare a well-organized speech but then fail to add the essential ingredient: practice. Constant practice distinguishes superior presentations from mediocre ones. It also helps to eliminate the nervousness that most speakers feel at one time or another.

In practicing your presentation, make use of four main techniques. They are listed here, from least effective to most effective:

- **Practice before a mirror:** This old-fashioned approach allows you to hear and see yourself in action. The drawback, of course, is that it is difficult to evaluate your own performance while you are speaking. Nevertheless, such run-throughs definitely make you more comfortable with the material.
- **Use of audiotape:** Most presenters have access to a tape player, so this approach is quite practical. The portability of the machines allows you to practice almost anywhere. Although taping a presentation will not im-

---

1. *Notes on Cards (3"× 5" or 4" × 6")*

Advantages

- Are easy to carry in a shirt pocket, coat, or purse
- Provide way to organize points, through ordering of cards
- Can lead to smooth delivery in that each card contains only one or two points that are easy to view
- Can be held in one hand, allowing you to move away from lectern while speaking

Disadvantages

- Keep you from viewing outline of entire speech
- Require that you flip through cards repeatedly in speech
- Can limit use of gestures with hands
- Can cause confusion if they are not in correct order

2. *Notes on Sheets of Paper*

Advantages

- Help you quickly view outline of entire speech
- Leave your hands free to use gestures
- Are less obvious than note cards, for no flipping is needed

Disadvantages

- Tend to tie you to lectern, where the sheets lie
- May cause slipups in delivery if you lose your place on the page

3. *Notes on Overhead Transparencies*

Advantages

- Introduce variety to audience
- Give visual reinforcement to audience
- Can be turned on and off

Disadvantages

- Cannot be altered as easily at last minute
- Must be presented neatly and in parallel style
- Involve the usual risks that reliance on machinery always introduces into your presentation

---

FIGURE 3–12
Speech notes

prove gestures, it will help you discover and eliminate verbal distractions such as filler words (*uhhhh, um, ya know*).

- **Use of live audience:** Groups of your colleagues, friends, or family—simulating a real audience—can provide the kinds of responses that approximate those of a real audience. In setting up this type of practice session, however, make certain that observers understand the criteria for a good presentation and are prepared to give an honest, forthright critique.

- **Use of videotape:** This practice technique allows you to see and hear yourself as others do. Your careful review of the tape, particularly when done with another qualified observer, can help you identify and eliminate problems with posture, eye contact, vocal patterns, and gestures. At first it can be a chilling experience, but soon you will get over the awkwardness of seeing yourself on film.

Speech Guideline 6: Speak Vigorously and Deliberately

"Vigorously" means with enthusiasm; "deliberately" means with care, attention, and appropriate emphasis on words and phrases. The importance of this guideline becomes clear when you think back to how you felt during the last speech you heard. At the very least, you expected the speaker to show interest in the subject and to demonstrate enthusiasm. Good information is not enough. You need to arouse the interest of the listeners.

You may wonder, "How much enthusiasm is enough?" The best way to answer this question is to hear or (preferably) watch yourself on tape. Your delivery should incorporate just enough enthusiasm so that it sounds and looks a bit unnatural to you. Few if any listeners ever complain about a speech being too enthusiastic or a speaker being too energetic. But many, many people complain about dull speakers who fail to show that they themselves are excited about the topic. Remember—every presentation is, in a sense, "show time."

Speech Guideline 7: Avoid Filler Words

Avoiding filler words presents a tremendous challenge to most speakers. When they think about what comes next or encounter a break in the speech, they may tend to fill the gap with filler words and phrases such as these:

> uhhhhh. . .
> ya know. . .
> okay. . .
> well. . . uh. . .
> like. . .
> I mean. . .
> umm. . .

These gap-fillers are a bit like spelling errors in written work: Once your listeners find a few, they start looking for more and are distracted from your presentation. To eliminate such distractions, follow these three steps:

Step 1: Use pauses to your advantage. Short gaps or pauses inform the listener that you are shifting from one point to another. In signaling a transition, a pause serves to draw attention to the point you make right after the pause. Note how listeners look at you when you pause. Do *not* fill these strategic pauses with filler words.

Step 2: Practice with tape. Tape is brutally honest: When you play it back, you will become instantly aware of fillers that occur more than once or twice. Keep a tally sheet of the fillers you use and their frequency. Your goal will be to reduce this frequency with every practice session.

Step 3: Ask for help from others. After working with tape machines in Step 2, give your speech to an individual who has been instructed to stop you after each filler. This technique gives immediate reinforcement.

Speech Guideline 8: Use Rhetorical Questions

Enthusiasm, of course, is your best delivery technique for capturing the attention of the audience. Another technique is the use of rhetorical questions at pivotal points in your presentation.

Rhetorical questions are those you ask to get listeners thinking about a topic, not those that you would expect them to answer out loud. They prod listeners to think about your point and set up an expectation that important information will follow. Also, they break the monotony of standard declarative sentence patterns. For example, here is a rhetorical question used by a computer salesperson in proposing a purchase:

> I've discussed the three main advantages that a centralized word processing center would provide your office staff. But is this an approach that you can afford at this point in the company's growth?

Then the speaker would follow the question with remarks supporting the position that the system is affordable.

"What if" scenarios provide another way to introduce rhetorical questions. They gain the listeners' attention by having them envision a situation that might occur. For example, a safety engineer could use this kind of rhetorical question in proposing asbestos-removal services to a regional bank:

> What if you repossessed a building that contained dangerous levels of asbestos? Do you think that your bank would then be liable for removing all the asbestos?

Again, the question pattern heightens listener interest.

Rhetorical questions do not come naturally. You must make a conscious effort to insert them at points when it is most important to gain or redirect the attention of the audience. Three particularly effective uses follow:

1. **As a grabber at the beginning of a speech:** "Have you ever wondered how you might improve the productivity of your word-processing staff?"

2. **As a transition between major points:** "We've seen that centralized word processing can improve the speed of report production, but will it require any additions to your staff?"
3. **As an attention-getter right before your conclusion:** "Now that we've examined the features of centralized word processing, what's the next step you should make?"

Speech Guideline 9: Maintain Eye Contact

Your main goal—always—is to keep listeners interested in what you are saying. This goal requires that you maintain control, using whatever techniques you can employ to direct the attention of the audience. Frequent eye contact is one good strategy.

The simple truth is that listeners pay closer attention to what you are saying when you look at them. Think how you react when a speaker makes constant eye contact with you. If you are like most people, you feel as if the speaker is speaking to you personally—even if there are 100 people in the audience. Also, you tend to feel more obligated to listen when you know that the speaker's eyes will be meeting yours throughout the presentation. Here are some ways you can make eye contact a natural part of your own strategy for effective oral presentations:

- **With audiences of about 30 or less:** Make regular eye contact with everyone in the room. Be particularly careful not to ignore members of the audience who are seated to your far right and far left (see Figure 3–13). Many speakers tend to focus on the listeners within Section B. Instead, make wide sweeps so that listeners in Sections A and C get equal attention.
- **With large audiences:** There may be too many people or a room too large for you to make individual eye contact with all listeners. In this case, focus on just a few people in all three sections of the audience noted in Figure 3–13. This approach gives the appearance that you are making eye contact with the entire audience.
- **With any size audience:** Occasionally look away from the audience— either to your notes or toward a part of the room where there are no faces looking back. In this way, you avoid the appearance of staring too intensely at your audience. Also, these breaks give you the chance to collect your thoughts or check your notes.

Speech Guideline 10: Use Appropriate Gestures and Posture

Speaking is only one part of giving a speech; another is adopting appropriate posture and using gestures that will reinforce what you are saying. Note that good speakers are much more than "talking heads" before a lectern. Instead, they:

1. Use their hands and fingers to emphasize major points
2. Stand straight, without leaning on or gripping the lectern

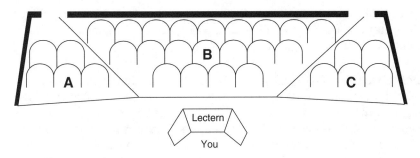

FIGURE 3–13
Audience sections

3. Step out from behind the lectern on occasion, to decrease the distance to the audience
4. Point toward visuals on screens or charts, without losing eye contact with the audience

The audience will judge you by what you say *and* what they see, a fact that again makes videotaping a crucial part of your preparation. With work on this facet of your presentation, you can avoid problems like keeping your hands constantly in your pockets, rustling change (remove pocket change and keys beforehand), tapping a pencil, scratching nervously, slouching over a lectern, and shifting from foot to foot.

■ GUIDELINES FOR SPEECH GRAPHICS

More than ever before, listeners expect good graphics during oral presentations. Much like gestures, graphics transform the words of your presentation into true communication with the audience. The following 10 guidelines will help you use graphics to enhance each speech.

Speech Graphics Guideline 1: Discover Listener Preferences

Some professionals prefer simple speech graphics, such as a conventional flip chart. Others prefer more sophisticated equipment, such as video projectors connected to laptop computers. For example, if your instructor in this course uses presentation graphics, he or she may have chosen overhead transparencies *or* a PowerPoint program.

Your listeners are usually willing to indicate their preferences when you call on them. Contact the audience ahead of time and make some inquiries. Also, ask for information about the room in which you will be speaking. If possible, request a setting that allows you to make best use of your graphics choice. If you have no control over the setting, then choose graphics that best fit the constraints. Details about lighting, wall space, and chair configuration can greatly influence your selection.

Speech Graphics Guideline 2: Think About Graphics Early

Graphics done as an afterthought usually look "tacked on." Plan graphics while you prepare the text so that the final presentation will seem fluid. This guideline holds true especially if you rely upon specialists to prepare your visuals. These professionals need some lead time to do their best work. Also, they can often provide helpful insights about how visuals will enhance the presentation—*if* you consult them early enough and *if* you make them a part of your presentation team.

The goal is to use graphics of which you can be proud. Never, never put yourself in the position of having to apologize for the quality of your graphic material. If an illustration is not up to the quality your audience would expect, do *not* use it.

Speech Graphics Guideline 3: Keep the Message Simple

Listeners may be suspicious of slick visual effects that appear to subordinate the content of the speech. Many prefer the simplicity of simple overhead transparencies and flip charts. However, you may prefer to use PowerPoint, video, or other high-tech graphics.

Speech Graphics Guideline 4: Make Any Wording Brief and Visible

Some basic design guidelines apply whether you are using posters, overhead transparencies, or computer-aided graphics such as PowerPoint.

- Use few words, emphasizing just one idea on each frame.
- Use much white space, perhaps as much as 60–70% per frame.
- Use "landscape" format more often than "portrait," especially since it is the preferred default setting for most presentation software.
- Use sans serif large print, from 14–18 pt. minimum for text to 48 pt. for titles.

Your goal should be to create graphics that are easily seen from anywhere in the room and that complement—but do *not* overpower—your presentation.

Speech Graphics Guideline 5: Use Colors Carefully

Colors can add flair to visuals. Follow these simple guidelines to make colors work for you:

- Have a good reason for using color (such as the need to highlight three different bars on a graph with three distinct colors).

■ Use only dark, easily seen colors, and be sure that a color contrasts with its background (for example, yellow on white would not work well).

■ Use no more than three or four colors in each graphic (to avoid a confused effect).

■ For variety, consider using white on a black or dark green background.

Speech Graphics Guideline 6:
Leave Graphics Up Long Enough

Because graphics reinforce text, they should be shown only while you address the particular point at hand. For example, reveal a graph just as you are saying, "As you can see from the graph, the projected revenue reaches a peak in 2005." Then pause and leave the graph up a bit longer for the audience to absorb your point.

How long is *too* long? A graphic outlives its usefulness when it remains in sight after you have moved on to another topic. Listeners will continue to study it and ignore what you are now saying. If you use a graphic once and plan to return to it, take it down after its first use and show it again later.

Speech Graphics Guideline 7: Avoid Handouts

Because timing is so important in your use of speech graphics, handouts are usually a bad idea. Readers move through a handout at their own pace, rather than at the pace the speaker might prefer. Thus handouts cause you to lose the attention of your audience. Use them only if (1) no other visual will do, (2) your listener has requested them, or (3) you distribute them as reference material *after* you have finished talking.

Speech Graphics Guideline 8: Maintain Eye
Contact While Using Graphics

Do not stare at your visuals while you speak. Maintain control of listeners' responses by looking back and forth from the visual to faces in the audience. To point to the graphic aid, use the hand closest to the visual. Using the opposite hand causes you to cross over your torso, forcing you to turn your neck and head away from the audience.

Speech Graphics Guideline 9: Include All Graphics in
Your Practice Sessions

Dry runs before the actual presentation should include every graphic you plan to use, in its final form. This is a good reason to prepare graphics as you prepare text, rather than as an afterthought. Running through a final practice without graphics would be much like doing a dress rehearsal for a play without costumes and props—you would be leaving out parts that require the greatest degree of timing and orchestration. Practicing with graphics helps you improve transitions.

Speech Graphics Guideline 10: Use Your Own Equipment

Murphy's Law always seems to apply when you use another person's audio-visual equipment: Whatever can go wrong, will. For example, a new bulb burns out, there is no extra bulb in the equipment drawer, an extension cord is too short, the screen does not stay down, the client's computer doesn't read your disk—many speakers have experienced these problems and more. Even if the equipment works, it often operates differently from what you are used to. The only sure way to put the odds in your favor is to carry your own equipment and set it up in advance.

However, most of us have to rely on someone else's equipment at least sometimes. Here are a few ways to ward off disaster:

- Find out exactly who will be responsible for providing the equipment and contact that person in advance.
- Have some easy-to-carry backup supplies in your car—an extension cord, an overhead projector bulb, felt-tip markers, and chalk, for example.
- Bring handout versions of your visuals to use as a last resort.

In short, you want to avoid putting yourself in the position of having to apologize. Plan well.

■ GUIDELINES FOR OVERCOMING NERVOUSNESS

The problem of nervousness deserves special mention because it is so common. Virtually everyone who gives speeches feels some degree of nervousness before "the event." An instinctive "fight or flight" response kicks in for the many people who have an absolute dread of presentations. As the cliché goes, do not try to eliminate "butterflies" before a presentation—just get them to fly in formation. It is best to acknowledge that a certain degree of nervousness will always remain. Then go about the business of getting it to work for you. Here are a few suggestions.

No Nerves Guideline 1: Know Your Speech

The most obvious suggestion is also the most important one. If you prepare your speech well, your command of the material will help to conquer any queasiness you feel—particularly at the beginning of the speech, when nervousness is usually at its peak. Be so sure of the material that your listeners will overlook any initial discomfort you may feel.

No Nerves Guideline 2: Prepare Yourself Physically

Your physical well-being before the speech can have a direct bearing on anxiety. More than ever before, most cultures understand the essential connection between mental and physical well-being. This connection suggests you should take these precautions before your presentation:

- **Avoid caffeine or alcohol for at least several hours before you speak.** You do not need the additional jitters brought on by caffeine or the false sense of ease brought on by alcohol.
- **Eat a light, well-balanced meal within a few hours of speaking.** However, do not overdo it—particularly if a meal comes right before your speech. If you are convinced that any eating will increase your anxiety, wait to eat until after speaking.
- **Practice deep-breathing exercises before you speak.** Inhale and exhale slowly, making your body slow down to a pace you can control. If you can control your breathing, you can probably keep the butterflies flying in formation.
- **Exercise normally the same day of the presentation.** A good walk will help invigorate you and reduce nervousness. However, do not wear yourself out by exercising more than you would normally.

No Nerves Guideline 3: Picture Yourself Giving a Great Presentation

Many speakers become nervous because their imaginations are working overtime. They envision the kinds of failure that almost never occur. Instead, speakers should be constantly bombarding their psyches with images of success, not failure. Mentally take yourself through the following steps of the presentation:

- Arriving at the room
- Feeling comfortable at your chair
- Getting encouraging looks from your audience
- Giving an attention-getting introduction
- Presenting your supporting points with clarity and smoothness
- Ending with an effective wrap-up
- Fielding questions with confidence

Sometimes called "imaging," this technique helps to program success into your thinking and to control negative feelings that pass through the minds of even the best speakers.

No Nerves Guideline 4: Arrange the Room as You Want

To control your anxiety, assert some control over the physical environment as well. You need everything going for you if you are to feel at ease. Make sure that chairs are arranged to your satisfaction, that the lectern is positioned to your taste, that the lighting is adequate, and so on. These features of the setting can almost always be adjusted if you make the effort to ask. Again, it is a matter of your asserting control to increase your overall confidence.

No Nerves Guideline 5: Have a Glass of Water Nearby
Extreme thirst and a dry throat are physical symptoms of nervousness that can affect delivery. There is nothing to worry about as long as you have water available. Think about this need ahead of time so that you do not have to interrupt your presentation to pour a glass of water.

No Nerves Guideline 6: Engage in Casual Banter Before the Speech
If you have the opportunity, chat with members of the audience before the speech. This ice-breaking technique will reduce your nervousness and help start your relationship with the audience.

No Nerves Guideline 7: Remember That You Are the Expert
As a final "psyching up" exercise before you speak, remind yourself that you have been invited or hired to speak on a topic about which you have useful knowledge. Your listeners want to hear what you have to say and are eager for you to provide useful information to them. So tell yourself, "I'm the expert here!"

No Nerves Guideline 8: Do Not Admit Nervousness to the Audience
No matter how anxious you may feel, never admit it to others. First of all, you do not want listeners to feel sorry for you—that is not an emotion that will lead to a positive critique of your speech. Second, nervousness is almost never apparent to the audience. Your heart may be pounding, your knees may be shaking, and your throat may be dry, but few if any members of the audience can see these symptoms. Why draw attention to the problem by admitting to it? Third, you can best defeat initial anxiety by simply pushing right on through.

No Nerves Guideline 9: Slow Down
Some speakers who feel nervous tend to speed through their presentations. If you have prepared well and practiced the speech on tape, you are not likely to let this happen. Having heard yourself on tape, you will be better able to sense that the pace is too quick. As you speak, constantly remind yourself to maintain an appropriate pace. If you have had this problem before, you might even write "Slow down!" in the margin of your notes.

No Nerves Guideline 10: Join a Speaking Organization
The previous nine guidelines will help reduce your anxiety about a particular speech. To help solve the problem over the long term, however, consider joining an organization like Toastmasters International, which promotes the speaking skills of all its members. Like some other speech organizations, Toastmasters has chapters that meet at many companies and campuses. These meetings provide an excellent, supportive environment in which all members can refine their speaking skills.

Appendix A:
ABC Formats and
Examples

*C*hapter 2 covered the ABC Format (Abstract/Body/Conclusion) in general. This appendix gets down to specifics by applying the ABC Format to 16 on-the-job documents. You can use this section as a desk reference while outlining and writing your draft.

Each of the 16 sections contains two parts: (1) an ABC outline for the document, along with some "Helpful Hints" at the bottom of the page, and (2) a document example that follows the outline. Documents are provided for the following ABC models:

1. Positive letter
2. Negative letter
3. Neutral letter
4. Sales letter
5. Memorandum
6. Instructions
7. Problem analysis
8. Recommendation report
9. Equipment evaluation
10. Lab report
11. Trip report
12. Feasibility study
13. Proposal
14. Progress/periodic report
15. Formal report
16. Job letter and resume

ABC Format 1: Positive Letter

Abstract

- Bridge between this letter and last communication with reader
- Clear statement of good news you have to report

Body

- Supporting data for main point mentioned in abstract
- Clarification of any questions reader may have
- Qualification, if any, of the good news

Conclusion

- Statement of eagerness to continue relationship, complete project, etc.
- Clear statement, if appropriate, of what step should come next

Helpful Hints

In the "bridge" (see Abstract), refer to your last contact with the reader or to the situation that prompted the letter (for example, "Thanks for your letter asking for a sample of our new Glide-Way bearings"). The good news comes immediately after the bridge—and sometimes even before it (for example, "Yes, we'll be glad to send you a free sample immediately"). In other words, good news always comes early in the letter. Any delay can cause confusion about the letter's purpose and can make the reader think, wrongly, that bad news is ahead.

Carruthers & Co.

12 Post Street
Houston, Texas 77000
(713) 555-9781

July 23, 2000

The Reverend John C. Davidson
Maxwell Street Church
Canyon Valley, Texas 79195

Dear Reverend Davidson:

Thanks for your letter asking to reschedule the church project from mid-August to another, more convenient time. Yes, we'll be able to do the project on one of two possible dates in September, as explained below.

As you know, Carruthers originally planned to fit your foundation investigation between two other projects planned for the Canyon Valley area. In making every effort to lessen church costs, we would be saving money by having a crew already on site in your area—rather than having to charge you mobilization costs to and from Canyon Valley.

As it happens, we have just agreed to perform another large project in the Canyon Valley area beginning on September 18. We would be glad to schedule your project either before or after that job. Specifically, we could be at the church site for our one-day field investigation on either September 17 or September 25, whatever date you prefer.

Please call me by September 2 to let me know your scheduling preference for the project. In the meantime, have a productive and enjoyable conference at the church next month.

Sincerely,

Nancy Slade

Nancy Slade, P.E.
Project Manager

NS/mh

EXAMPLE 1
Positive letter
Source: Technical Writing: A Practical Approach, 4th ed. (p. 242) by W. S. Pfeiffer, 2000, Upper Saddle River, NJ: Prentice Hall. Reprinted by permission.

ABC Format 2: Negative Letter

Abstract

- Bridge between your letter and previous communication
- General statement of purpose or appreciation—in an effort to find common bond or area of agreement

Body

- Strong emphasis on what can be done, when possible
- Buffered (yet clear) statement of what cannot be done, with reasons
- Facts that support your views

Conclusion

- Closing remarks that express interest in continued association
- Statement, if appropriate, of what will happen next

Helpful Hints

In negative letters, buffer the bad news but still be clear. Use the abstract to convey information that will help preserve good will and your relationship with the reader. Then use the body of the letter to state what can—and what cannot—be done. This approach assumes you want to retain your relationship with the reader and avoid "burning bridges." In the few cases where the relationship with the reader is gone—such as a request for payment that has been repeatedly ignored—a more direct approach may be necessary.

12 Post Street
Houston, Texas 77000
(713) 555-9781

July 23, 2000

The Reverend John C. Davidson
Maxwell Street Church
Canyon Valley, Texas 79195

Dear Reverend Davidson:

Thanks for your letter asking to reschedule the foundation project at your church from mid-August to late August, because of the regional conference. I am sure you are proud that Maxwell was chosen as the conference site.

One reason for our original schedule, as you may recall, was to save the travel costs for a project crew going back and forth between Houston and Canyon Valley. Because Carruthers has several other jobs in the area, we had planned not to charge you for travel.

We can reschedule the project, as you request, to a more convenient date in late August, but the change will increase project costs from $1,500 to $1,800 to cover travel. At this point, we just don't have any other projects scheduled in your area in late August that would help defray the additional expenses. Given our low profit margin on such jobs, that additional $300 would make the difference between our firm making or losing money on the foundation investigation at your church.

I'll call you next week, Reverend Davidson, to select a new date that would be most suitable. Carruthers welcomes its association with the Maxwell Street Church and looks forward to a successful project in late August.

Sincerely,

Nancy Slade

Nancy Slade, P.E.
Project Manager

NS/mh

EXAMPLE 2
Negative letter
Source: Technical Writing: A Practical Approach, 4th ed. (p. 243) by W. S. Pfeiffer, 2000, Upper Saddle River, NJ: Prentice Hall. Reprinted by permission.

ABC Format 3: Neutral Letter

Abstract

- Bridge or transition between letter and previous communication
- Exact purpose of letter (request, invitation, etc.)

Body

- Details that support purpose statement—for example,
  - —Description of item(s) requested
  - —Requirements related to the invitation
  - —Description of item(s) sent

Conclusion

- Statement of appreciation
- Description of actions that should occur next

Helpful Hints

Neutral letters must be absolutely clear about your inquiry or response. As in positive letters, make your point early so that there is no confusion about the message. Your first paragraph can include a bridge or transition sentence (for example, "Yesterday I called your warehouse to find out what calculators you have in stock"). It can be followed by a direct reference to the request (for example, "Using information provided by your staff, I want to order the nine calculators listed below").

River
College

January 4, 2000

Mr. Timothy Fu, Personnel Director
The Franklin Group
127 Rainbow Lane
St. Louis, MO 63103

Dear Mr. Fu:

The Franklin Group has hired 35 graduates of River College since 1975. To help continue that tradition, we would like to invite you to the college's first Career Fair, to be held February 21, 2000, from 8 a.m. until noon.

Sponsored by the Student Government Association, the Career Fair gives juniors and seniors the opportunity to get to know more about a number of potential employers. We give special attention to organizations, like The Franklin Group, that have already had success in hiring River College graduates. Indeed, we have already had a number of inquiries about whether your firm will be represented at the fair.

Participating in the Career Fair is simple. We will provide you with a booth where one or two Franklin representatives can talk with students who come by to ask about your firm's career opportunities. Feel free to bring along whatever brochures or other written information that would help our students learn more about Franklin's products and services.

I will call you next week, Mr. Fu, to give more details about the fair and to offer a specific booth location. We at River College look forward to building on our already strong association with The Franklin Group.

Sincerely,

Faron G. Abdullah

Faron G. Abdullah, President
Student Government Association

56 New Lane
Bolt, Missouri
65101
(314) 555-0272

EXAMPLE 3
Neutral letter
Source: Adapted from *Technical Writing: A Practical Approach,* 4th ed. (p. 244) by W. S. Pfeiffer, 2000, Upper Saddle River, NJ: Prentice Hall. Reprinted by permission.

ABC Format 4: Sales Letter

Abstract

- Choose one or two of the following strategies for capturing attention:
  —Cite a surprising fact
  —Announce a new product or service
  —Ask a question
  —Show understanding of client's problem
  —Show potential for solving client's problem
  —Summarize results of a meeting
  —Answer a question reader previously asked

Body

- Choose one or two of the following strategies for convincing the reader:
  —Stress one main problem reader has concern about
  —Stress one main selling point of your solution
  —Emphasize what is unique about your solution
  —Focus on value and quality, rather than price
  —Put details in enclosures
  —Briefly explain the value of any enclosures

Conclusion

- Keep control of the next step by doing the following:
  —Leave the reader with one crucial point to remember
  —Offer to call (first choice) or ask reader to call (last choice)

Helpful Hints

The key words above are capture, convince, and control—the "3Cs." Capture the readers' attention at the beginning, convince their minds in the middle, and control the next step at the end. Sales letters work together with personal contacts to build continuing relationships with customers.

MASTMAN SAFETY RESEARCH CONSULTANTS, Inc.

August 21, 2000

Mr. James Swartz, Safety Director
Jessup County School System
1111 Clay Street
Smiley, MO 64607

NEW ASBESTOS-ABATEMENT SERVICE NOW AVAILABLE

We enjoyed working with you last year, James, to update your entire fire alarm system. Given the current concern in the country about another safety issue, asbestos, we wanted you to know that our staff now does abatement work.

As you know, many of the state's school buildings were constructed during years when asbestos was used as a primary insulator. No one knew then, of course, that the material can cause illness and even premature death for those who work in buildings where asbestos was used in construction. Now we know that just a small portion of asbestos produces a major health hazard.

Fortunately, there's a way to tell whether you have a problem: the asbestos survey. This procedure, done by our certified asbestos-abatement professionals, results in a report that tells whether your buildings are affected. If we find asbestos, we can remove it for you.

Jessup showed foresight in modernizing its alarm system last year, James. Your desire for a thorough job on that project was matched, as you know, by the approach we take to our business. Now we'd like to give you the peace of mind that will come from knowing that either (1) there is no asbestos problem in your 35 structures or (2) you have removed the material.

The enclosed brochure outlines our asbestos services. I'll call you in a few days to see how Mastman can help you.

Barbara Feinstein

Barbara H. Feinstein
Certified Industrial Hygienist

EXAMPLE 4
Sales letter
Source: Adapted from *Technical Writing: A Practical Approach,* 4th ed. (p. 246) by W. S. Pfeiffer, 2000, Upper Saddle River, NJ: Prentice Hall. Reprinted by permission.

ABC Format 5: Memorandum

Abstract

- Clear statement of memo's purpose
- Outline of main parts of memo

Body

- Absolute clarity about what memo has to do with reader
- Tactful presentation of any negative news
- Supporting points, with strong points at the beginning or end
- Frequent use of short paragraphs or lists
- Reference to attachments, when much detail is required

Conclusion

- Clear statement of what step should occur next
- An effort to retain goodwill and cooperation of readers

Helpful Hints

We spend more time reading and writing memos (including email) than we want to admit. The main rule is to be clear, brief, and tactful. Because many activities are competing for your readers' time, memos should be written so that they can be understood immediately. Use headings and lists. Also, put the main point at the beginning, with a brief buffer if you are delivering bad news.

Totally Teak, Inc.

MEMORANDUM

DATE: August 1, 2000
TO: Technical Staff
FROM: Gini Preston, Chair, Word Processing Committee *GP*
SUBJECT: Word Processing Suggestions

The Word Processing Committee has met for six weeks to consider changes in Totally Teak's Word Processing Center. This memo highlights the recommendations that have been approved by management.

1. **Document status:** Documents will be designated either "rush" or "regular" status, depending on what you request. If at all possible, rush documents will be returned within four hours. Regular documents will be returned within one working day.

2. **Draft stages:** Both users and operators should make every effort to produce no more than three hard-copy drafts of any document. Typically, these would include the following:

   - **First typed draft** (typed from writer's handwritten or cut-and-paste copy)
   - **Second typed draft** (produced after user has made editing corrections on first-draft copy)
   - **Final typed draft** (produced after user makes final editing changes, after the proofreader makes a pass through the document, and after the operator incorporates final changes into the copy)

3. **New proofreader:** A company proofreader has been hired to improve the quality of our documents. This individual will have an office in the Word Processing Center and will review all documents produced by the word processing operators.

These changes will take effect August 15. Your efforts to implement them will help improve the efficiency of the center, the quality of your documents, and the productivity of the company.

Feel free to call me at ext. 567 if you have any questions.

Copy: Rob Ahab

EXAMPLE 5
Memorandum
Source: Adapted from *Technical Writing: A Practical Approach,* 4th ed. (p. 247) by W. S. Pfeiffer, 2000, Upper Saddle River, NJ: Prentice Hall. Reprinted by permission.

ABC Format 6: Instructions

Abstract

■ Clear purpose statement for instructions
■ Summary of main steps
■ List of materials or equipment needed (or reference to list or graphic)

Body

■ Helpful pointers or definitions
■ Well-placed references to (1) Cautions (possibility of damage to things), (2) Warnings (possibility of injury), and (3) Dangers (probability of injury or death)
■ Numbered steps
　—Steps grouped under main tasks, not presented in one "laundry list"
　—Limit of one action in each step
　—Use of verbs rather than nouns ("Check the meter reading")
■ Separation of essential information (in steps) from helpful information (in "Notes" and "Results")

Conclusion

■ Statement, or restatement, about importance of procedure

Helpful Hints

Instructions can be either freestanding documents or part of another document. In either case, the most common error is to make them too complicated for the audience. Carefully consider the technical level of your readers. Use white space, graphics, and other design elements to make the instructions appealing. Most important, be sure to include Caution, Warning, and Danger references *before* the steps to which they apply.

GENERAL
CONSULTING
CONTRACTORS

MEMORANDUM

DATE: June 23, 2000
TO: Employees Receiving New Message Recorders
FROM: Matthew Edwards, Purchasing Agent *ME*
SUBJECT: Instructions for New Message Recorders

INTRODUCTORY SUMMARY

We have just received the new phone message recorder you ordered. After processing, it will be delivered to your office within the week. The machine is one of the best on the market, but the instructions that accompany it are somewhat hard to follow. To help you begin using the recorder as soon as possible, I have simplified the instructions for setting up and operating the machine.

The illustration below labels the machine's parts. Following the illustration are seven easy steps you need to operate the recorder.

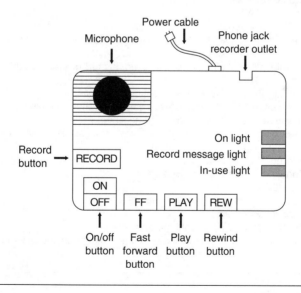

EXAMPLE 6 (pp. 77–79)
Instructions
Source: Adapted from *Technical Writing: A Practical Approach,* 4th ed. (pp. 199–201) by W. S. Pfeiffer, 2000, Upper Saddle River, NJ: Prentice Hall. Reprinted by permission.

Memo to: Employees receiving new message recorders   Page 2
June 23, 2000

SETTING UP AND USING YOUR NEW RECORDER
If you devote about 15 minutes to these 7 tasks, you can learn to operate your new message recorder.

1. Hooking Up Your Recorder
   a. Plug the recorder into any wall outlet using the *power cable.*
   b. Plug the *phone jack* into the *recorder outlet* located at the back of the unit.

2. Preparing Your Message
   a. Write down your:
      Greeting
      Name and department
      Time of return
   b. Write down the request you want to leave on the machine.
      NOTE: A sample request might be as follows: "Please leave your name, phone number, and a brief message after you hear the tone."

3. Recording Your Message
   a. Press the *ON* button.
      RESULT: The red *on* light will come on.
   b. Press and hold down the *RECORD* button, and keep holding it down for the entire time you record.
      RESULT: While the button is held down, the red *record message* light will come on.
   c. Record your message directly into the *microphone.*
   d. Release the *RECORD* button when you are finished recording.
   e. Do you need to record the message again?
      If *yes,* repeat steps a through d (your previous message will be erased each time you record).
      If *no,* go on to the next step.

4. Turning On and Testing Your Recorder
   a. Press the *ON* button.
      RESULT: The red *on* light will come on.
   b. Press the *PLAY* button.
      RESULT: The red *in use* light will come on.
   c. Call in your message from another phone to make sure the unit is recording properly.

EXAMPLE 6 (*continued*)

5. Playing Back Messages

a. Look at the red *record message* light to see if it is blinking.
   NOTE: The number of times the light blinks in succession indicates the number of messages you have received.
b. Press the *REW* (rewind) button.
   NOTE: The tape will stop automatically when it is completely rewound.

> CAUTION: Do not press *PLAY* and *FF* (fast forward) at the same time! Doing so will break the tape. See the manufacturer's manual for process of replacing broken tape.

c. Press *PLAY* and listen to the messages.
d. Do you want to replay messages?
   If *yes*, repeat steps b and c.
   If *no*, go on to next step.
e. Do you want to skip ahead to other messages?
   If *yes*, push the *FF* (fast forward) button.
   If *no*, go on to next step.

6. Erasing Received Messages

a. Press and continue holding down the *PLAY* and *REW* (rewind) buttons at the same time.
b. Release the buttons when you hear a click.
   NOTE: The tape automatically stops when the messages are erased.

7. Turning OFF Your Recorder

a. Press the *OFF* button.
   RESULT: All red lights will go off.

CONCLUSION

These new recorders are fully guaranteed for three years, so please report any problems right away. Paul Hansey (ext. 765) will be glad to help fix the machine or return the machine to the manufacturer for repair. In particular, you need to report these problems to Paul:

- Lost or incomplete messages
- Interference or noise on line
- Faulty equipment
- Inability to record

ABC Format 7: Problem Analysis

Abstract

- Purpose of report
- Capsule summary of problems covered in report discussion

Body

- Background on source of problems
- Well-organized description of problems observed
- Data that support your observations
- Consequences of the problems

Conclusion

- Brief restatement of main problems (unless report is so short that restatement would be repetitious)
- Degree of urgency required in handling problems
- Suggested next step

Helpful Hints

Problem analyses must provide an *objective* presentation of information, so that the readers (who are often decision-makers) can decide upon the next step. Any opinions set forth must be well supported by facts.

Totally Teak, Inc.

MEMORANDUM

DATE: October 15, 2000
TO: Jan Stillwright, Vice President of Research and Training
FROM: Harold Marshal, Technical Supervisor *HM*
SUBJECT: Boat Problems During Summer Season

INTRODUCTORY SUMMARY

We have just completed a one-month project in the Pacific Ocean aboard the leased ship, *Seeker II*. All work went just about as planned, with very few delays caused by weather or equipment failure.

However, there were some boat problems that need to be solved before we lease *Seeker II* again this season. This report highlights the problems so that they can be brought to the owner's attention. My comments focus on four areas of the boat: drill rig, engineering lab, main engine, and crew quarters.

DRILL RIG

Thus far the rig has operated without incident. Yet on one occasion, I noticed that the elevator for lifting pipe up the derrick swung too close to the derrick itself. A quick gust of wind or a sudden increase in sea height caused these shifts. If the elevator were to hit the derrick, causing the elevator door to open, pipe sections might fall to the deck below.

I believe the whole rig assembly must be checked by someone knowledgeable about its design. Before we put workers near that rig again, we need to know that their safety would not be jeopardized by the possibility of falling pipe.

ENGINEERING LAB

Quite frankly, it is a tribute to our technicians that they were able to complete all lab tests with *Seeker II*'s limited facilities. Several weeks into the voyage, four main problems became apparent:

1. Ceiling leaks
2. Poor water pressure in the cleanup sink
3. Leaks around the window near the electronics corner
4. Two broken outlet plugs

Although we were able to devise a solution to the window leaks, the other problems stayed with us for the entire trip.

EXAMPLE 7 (pp. 81–82)

Problem analysis

Source: Adapted from *Technical Writing: A Practical Approach,* 4th ed. (pp. 278–279) by W. S. Pfeiffer, 2000, Upper Saddle River, NJ: Prentice Hall. Reprinted by permission.

MAIN ENGINE

On this trip, we had three valve failures on three different cylinder heads. From our experience on other ships, it is very unusual to have one valve fail, let alone three. Fortunately for us, these failures occurred between projects, so we did not lose time on a job. And fortunately for the owner, the broken valve parts did not destroy the engine's expensive turbocharger.

Only an expert will be able to tell whether these engine problems were flukes or if the entire motor needs to be rebuilt. In my opinion, the most prudent course of action is to have the engine checked over carefully before the next voyage.

CREW QUARTERS

When 15 men live in one room for three months, it is important that basic facilities work. On *Seeker II* we experienced problems with the bedroom, bathroom, and laundry room that caused some tension.

Bedroom

Three of the top bunks had such poor springs that the occupants sank 6 to 12 in. toward the bottom bunks. More important, five of the bunks are not structurally sound enough to keep from swaying in medium to high seas. Finally, most of the locker handles are either broken or about to break.

Bathroom

Poor pressure in three of the commodes made them almost unusable during the last two weeks. Our amateur repairs did not solve the problem, so I think the plumbing leading to the holding tank might be defective.

Laundry Room

We discovered early that the filtering system could not screen the large amount of rust in the old 10,000-gallon tank. Consequently, undergarments and other white clothes turned a yellow-red color and were ruined.

CONCLUSION

As noted at the outset, none of these problems kept us from accomplishing the major goals of this voyage. But they did make the trip much more uncomfortable than it had to be. Moreover, in the case of the rig and engine problems, we were fortunate that injuries and downtime did not occur.

I strongly urge that the owner be asked to correct these deficiencies before we consider using *Seeker II* for additional projects this season.

EXAMPLE 7 (*continued*)

ABC Format 8: Recommendation Report

Abstract

- Purpose of report
- Brief reference to problem to which recommendations respond
- Capsule summary of recommendations covered in report

Body

- Details about problem
- Well-organized description of recommendations
- Data that support recommendations (with reference to attachments)
- Main benefits of recommendations
- Any possible drawbacks

Conclusion

- Brief restatement of main recommendations
- Main benefit of recommended change
- Your offer to help with next step

Helpful Hints

Recommendation reports must be more persuasive than reports like problem analyses. Yet any recommendations in your report must be well supported by facts and analysis. You want readers to see your recommendations as ideas that flow naturally and inevitably from facts in the report.

GCC

GENERAL
CONSULTING
CONTRACTORS

677 Rothrock Way
Fairfax, Virginia 22030
(703) 555-6273

April 22, 2000

Big Muddy Oil Company, Inc.
12 Rankin St.
Abilene, TX 79224

ATTENTION: Mr. James Smith, Engineering Manager

SHARK PASS STUDY
BLOCK 15, AREA 43-B, GULF OF MEXICO

INTRODUCTORY SUMMARY

You recently asked our firm to complete a preliminary soils investigation at an offshore rig site. This report presents the tentative results of our study, including major conclusions and recommendations. A longer, formal report will follow at the end of the project.

On the basis of what we have learned so far, we believe that you can safely place an oil platform at Shark Pass site. To limit the chance of a rig leg punching into the seafloor, however, we suggest you follow the recommendations in this report.

WORK AT THE PROJECT SITE

On April 16 and 17, 2000, GCC's engineers and technicians worked at the Block 15 site in the Shark Pass region of the gulf. Using GCC's leased drill ship, *Atlantis,* as a base of operations, our crew performed these main tasks:

- Seismic survey of the project study area
- Two soil borings of 40 feet each

Both seismic data and soil samples were brought to our Houston office for laboratory analysis.

LABORATORY ANALYSIS

On April 18 and 19, our lab staff examined the soil samples, completed bearing capacity tests, and evaluated seismic data. Here are the results of that analysis.

Soil Layers

Our initial evaluation of the soil samples reveals a 7–9 ft layer of weak clay starting a few feet below the seafloor. Other than that layer, the composition of the soils seems fairly typical of other sites nearby.

EXAMPLE 8 (pp. 84–85)

Recommendation report

Source: Technical Writing: A Practical Approach, 4th ed. (pp. 274–275) by W. S. Pfeiffer, 2000, Upper Saddle River, NJ: Prentice Hall. Reprinted by permission.

James Smith   Page 2
April 22, 2000

Bearing Capacity

We used the most reliable procedure available, the XYZ method, to determine the soil's bearing capacity (that is, its ability to withstand the weight of a loaded oil rig). That method required that we apply the following formula:

Q = $cN_v + tY$, where
Q = ultimate bearing capacity
c = average cohesive shear strength
N_v = the dimensionless bearing capacity factor
t = footing displacement
Y = weight of the soil unit

The final bearing capacity figure will be submitted in the final report, after we repeat the tests.

Seafloor Surface

By pulling our underwater seismometer back and forth across the project site, we developed a seismic "map" of the seafloor surface. That map seems typical of the flat floor expected in that area of the gulf. The only exception is the presence of what appears to be a small sunken boat. This wreck, however, is not in the immediate area of the proposed platform site.

CONCLUSIONS AND RECOMMENDATIONS

On the basis of our analysis, we conclude that there is only a slight risk of instability at the site. Though unlikely, it is possible that a rig leg could punch through the seafloor, either during or after loading. We base this opinion on (1) the existence of the weak clay layer, and (2) the marginal bearing capacity.

Nevertheless, we believe you can still place your platform if you follow careful rig-loading procedures. Specifically, take these precautions to reduce your risk:

1. Load the rig in 10-ton increments, waiting one hour between loadings.
2. Allow the rig to stand 24 hours after the loading and before placement of workers on board.
3. Have a soils specialist observe the entire loading process, to assist with any emergency decisions if problems arise.

As noted at the outset, these conclusions and recommendations are based on preliminary data and analysis. We will complete our final study in three weeks and submit a formal report shortly thereafter.

GCC enjoyed working once again for Big Muddy Oil at its Gulf of Mexico lease holdings. I will phone you this week to see if you have any questions about our study. If you need information before then, please give me a call.

Sincerely,

Bartley Hopkins

Bartley Hopkins, Project Manager

ABC Format 9: Equipment Evaluation

Abstract

- Purpose of report
- Capsule summary of what report says about the equipment

Body

- Thorough description of equipment being evaluated
- Well-organized critique, either analyzing the parts of one piece of equipment or contrasting several pieces of similar equipment
- Additional supporting data, with reference to attachments

Conclusion

- Brief restatement of major findings, conclusions, or recommendations

Helpful Hints

Equipment evaluations give objective critiques about how equipment has functioned. Possible topics include machinery, tools, vehicles, software, and office supplies. Like a problem analysis, the equipment evaluation may focus on problems. Or, like a recommendation report, it may suggest a change. In any case, it must provide well-documented observations of the manner in which equipment has performed.

MEMORANDUM

DATE: July 26, 2000
TO: Melanie Frank, Office Manager
FROM: Hank Worley, Project Manager *HW*
SUBJECT: Evaluation of Best Choice Software

INTRODUCTORY SUMMARY
When the office purchased one copy of Best Choice Software last month, you suggested I send you an evaluation after 30 days' use. Having now used Best Choice for a month, I have concluded that it meets all our performance expectations. This memo presents our evaluation of the main features of Best Choice.

HOW BEST CHOICE HAS HELPED US
Best Choice provides five primary features: word processing, file management, spreadsheet, graphics, and a user's guide. Here is my critique of all five.

Word Processing
The system contains an excellent word-processing package that the engineers as well as the secretaries have been able to learn easily. This package can handle both our routine correspondence and the lengthy reports that our group generates. Of particular help is the system's 90,000-word dictionary, which can be updated at any time. The spelling correction feature has already saved much effort that was previously devoted to mechanical editing.

File Management
The file-manager function allows the user to enter information and then to manipulate it quickly. During one three-day site visit, for example, a field engineer recorded a series of problems observed in the field. Then she rearranged the data to highlight specific points I asked her to study, such as I-beam welds and concrete cracks.

Spreadsheet
Like the system's word-processing package, the spreadsheet is efficient and quickly learned. Because Best Choice is a multipurpose software package, spreadsheet data can be incorporated into letter or report format. In other words, spreadsheet information can be merged with our document format to create a final draft for submission to clients or supervisors, with a real savings in time. For example, the memo I sent you last week on budget projections for field equipment took me only an hour to complete; last quarter, the identical project took four hours.

EXAMPLE 9 (pp. 87–88)
Equipment evaluation
Source: Technical Writing: A Practical Approach, 4th ed. (pp. 276–277) by W. S. Pfeiffer, 2000, Upper Saddle River, NJ: Prentice Hall. Reprinted by permission.

Graphics

The graphics package permits visuals to be drawn from the data contained in the spreadsheet. For example, a pie chart that shows the breakdown of a project budget can be created easily by merging spreadsheet data with the graphics software. With visuals becoming such an important part of reports, we have used this feature of Best Choice quite frequently.

User's Guide

Eight employees in my group have now used the Best Choice user's guide. All have found it well laid out and thorough. Perhaps the best indication of this fact is that in 30 days of daily use, we have placed only three calls to the Best Choice customer-service number.

CONCLUSION

Best Choice seems to contain just the right combination of tools to help us do our job, both in the field and in the office. These are the system's main benefits:

- Versatility—it has diverse functions
- Simplicity—it is easy to master

The people in our group have been very pleased with the package during this 30-day trial. If you like, we would be glad to evaluate Best Choice for a longer period.

EXAMPLE 9 (*continued*)

ABC Format 10: Lab Report

Abstract

- Purpose of report
- Capsule summary of results

Body

- Purpose or hypothesis of lab work
- Equipment needed
- Procedures or methods used in the lab test
- Unusual problems or occurrences
- Results of the test with reference to your expectations (results may appear in conclusion, instead)

Conclusion

- Restatement of main results
- Implications of lab test for further work

Helpful Hints

Like procedures, lab reports may stand on their own or be part of a larger report that uses lab work for supporting detail. Lab reports usually include topics such as procedures, equipment, problems, results, and implications.

arruthers & Co.

12 Post Street
Houston, Texas 77000
(713) 555-9781

December 12, 2000

Mr. Andrew Hawkes
Monson Coal Company
2139 Lasiter Dr.
Baltimore, MD 21222

LABORATORY REPORT
BOREHOLE FOSSIL SAMPLES
BRAINTREE CREEK SITE, WEST VIRGINIA

INTRODUCTORY SUMMARY

Last week you sent us six fossil samples from the Braintree Creek site. Having analyzed the samples in our lab, we believe they suggest the presence of coal-bearing rock. As you requested, this report will give a summary of the materials and procedures we used in this project, along with any problems we had.

As you know, our methodology is to identify microfossils in the samples, estimate the age of the rock by when the microfossils existed, and then make assumptions about whether the surrounding rock might contain coal.

LAB MATERIALS

Our lab analysis relies on only one piece of specialized equipment: a Piketon electron microscope. Besides the Piketon, we use a simple 400-power manual microscope. Other equipment is similar to that included in any basic geology lab, such as filtering screens and burners.

LAB PROCEDURE

Once we receive a sample, we first try to identify the kinds of microfossils the rocks contain. Our specific lab procedure for your samples consisted of two steps:

Step 1

We used a 400-power microscope to visually classify the microfossils that were present. Upon inspection of the samples, we concluded that there were two main types of microfossils: nannoplankton and foraminifera.

EXAMPLE 10 (pp. 90–91)
Lab report
Source: Adapted from *Technical Writing: A Practical Approach,* 4th ed. (pp. 283–284) by W. S. Pfeiffer, 2000, Upper Saddle River, NJ: Prentice Hall. Reprinted by permission.

Step 2

Next, we had to extract the microfossils from the core samples you provided. We used two different techniques:

Nannoplankton Extraction Technique

a. Selected a pebble-size piece of the sample
b. Thoroughly crushed the piece under water
c. Used a dropper to remove some of the material that floats to the surface (it contains the nannoplankton)
d. Dried the nannoplankton-water combination
e. Placed the nannoplankton on a slide

Foraminifera Extraction Technique

a. Boiled a small portion of the sample
b. Used a microscreen to remove clay and other unwanted material
c. Dried remaining material (foraminifera)
d. Placed foraminifera on slide

PROBLEMS ENCOUNTERED

The entire lab procedure went as planned. The only problem was minor and occurred when we removed one of the samples from its shipping container. As the bag was taken from the shipping box, it broke open. The sample shattered when it fell onto the table. Fortunately, we had an extra sample from the same location.

CONCLUSION

Judging by the types of fossils present in the sample, we believe they come from rock of an age that might contain coal. This conclusion is based on limited testing, so we suggest you test more samples at the site. We would be glad to help with additional sampling and testing.

I will call you this week to discuss our study and any possible follow-up you may wish us to do.

Sincerely,

Joseph Rappaport

Joseph Rappaport
Senior Geologist

ABC Format 11: Trip Report

Abstract

- Destination, dates, and purpose of trip
- Brief overview of results of trip
- Main sections of report

Body

- Major accomplishments, grouped by task
- Any conclusions that flow from your work
- Any recommendations you developed
- Administrative details, such as trip expenses

Conclusion

- Wrap-up that refers to overall usefulness of the trip
- Follow-up activities that may be required

Helpful Hints

Trip reports can be so routine at some organizations that brief forms are used. In other organizations they are anything but routine. Assume the latter unless you know otherwise. Your organization has invested in your trip and expects you to be able to describe your accomplishments.

MASTMAN SAFETY RESEARCH CONSULTANTS, Inc.

MEMORANDUM

DATE: July 15, 2000
TO: Susan Newton, Manager of Marketing
FROM: Stone Prentice, Marketing Specialist *SP*
SUBJECT: Report on Trip to Seattle

Last Friday I visited Seattle to explore several marketing opportunities for our firm. Having met with two potential clients, I believe we should strongly consider entering the Seattle market. This report highlights the results of my trip and proposes some follow-up activities for your review.

MEETINGS WITH POTENTIAL CLIENTS
In Seattle I met with officials from the two companies you asked me to contact. Both expressed interest in Mastman's services, as noted below.

Meeting with Josh McDonald, Pacific Retro Services
I spent the morning of July 11 with Mr. McDonald, reviewing his firm's projects for the next two years. He wants us to submit proposals for three upcoming projects to improve the indoor air quality (IAQ) of several downtown buildings. Further, he noted he has heard good comments about our patented IAQ equipment.

Meeting with Maureen Hemphill, the Builders Group
In the afternoon of July 11, I met with Ms. Hemphill at one of her firm's largest current projects, the 110 Jackson Building. She noted that her executive staff has been unhappy with delays in construction; it would like to consider hiring a firm like ours to manage construction for future projects. In fact, the Builders Group will issue two large requests for proposals before September.

PROPOSED PLAN OF ACTION
Pacific Retro Services and the Builders Group may offer us a lucrative Seattle market for our services. I suggest a threefold plan of action:

1. Submit proposals for all future projects for which we are qualified
2. Study the feasibility of starting a small branch in Seattle
3. Conduct future marketing trips in Seattle

CLOSING
I would be glad to help with the follow-up activities listed above. Seattle offers us a good opportunity to expand our work and provide a useful service.

EXAMPLE 11
Trip report

ABC Format 12: Feasibility Study

Abstract

- Brief statement about who authorized study and for what purpose
- Brief statement of overall recommendation
- Reference to main parts of report

Body

- Description of methods used in study
- Description of evaluation criteria (cost, schedule, quality, etc.)
- Analysis of item or items, according to the evaluation criteria
- Detailed information about advantages and disadvantages of adopting the change
- Other alternatives to consider, if any

Conclusion

- Major conclusions and recommendations resulting from study
- Follow-up tasks that may be required to acquire additional data

Helpful Hints

Unlike proposals, feasibility studies are always requested by the readers. They are needed to determine the practicality—that is, feasibility—of a proposed policy, product, or service. Readers need information from the study to make decisions.

MEMORANDUM

DATE: July 22, 2000
TO: Greg Bass
FROM: Mike Tran *MT*
SUBJECT: Replacement of In-House File Server

INTRODUCTORY SUMMARY

The purpose of this feasibility study is to determine if the NTR PC905 would make a practical replacement for our in-house file server. As we agreed in our weekly staff meeting, our current file-serving computer is damaged beyond repair and must be replaced by the end of the week. This study shows that the NTR PC905 is a suitable replacement that we can purchase within our budget and install by Friday afternoon.

FEASIBILITY CRITERIA

There are three major criteria that I addressed. First, the computer we buy must be able to perform the tasks of a file-serving computer on our in-house network. Second, it must be priced within our $4,000 budget for the project. Third, it must be delivered and installed by Friday afternoon.

Performance

As a file server, the computer we buy must be able to satisfy these criteria:
- Store all programs used by network computers
- Store the source code and customer-specific files for Xtracheck
- Provide fast transfer of files between computers while serving as host to the network
- Serve as the printing station for the network laser printer

The NTR PC905 comes with a 120MB hard drive. This capacity will provide an adequate amount of storage for all programs that will reside on the file server. Our requirements are for 30MB of storage for programs used by network computers and 35MB of storage for source code and customer-specific programs. The 120MB drive will leave us with 55MB of storage for future growth.

The PC905 can transfer files and execute programs across our network. It can do so at speeds up to five times faster than our current file server. Productivity should increase because of less time spent waiting for transfer.

EXAMPLE 12 (pp. 95–96)

Feasibility study

Source: Adapted from *Technical Writing: A Practical Approach,* 4th ed. (pp. 393–394) by W. S. Pfeiffer, 2000, Upper Saddle River, NJ: Prentice Hall. Reprinted by permission.

Greg Bass
July 22, 2000
Page 2

The computer we choose as the file server must also serve as the printing station for our network laser printer. The PC905 is compatible with our Hewy Packer laser printer. It also has 2.0MB more memory than our current server. As a result, it can store larger documents in memory and print with greater speed.

Budget
The budget for the new file server is $4,000. The cost of the PC905 is as follows:

PC905 with 120MB Hard Drive	$2,910
Keyboard	112
Monitor	159
Total	$3,181

No new network boards need to be purchased because we can use those that are in the current server. We also have all additional hardware and cables that will be required for installation. Thus the PC905 can be purchased for $800 under budget.

Time Frame
Our sales representative at NTR guarantees delivery of the system by Friday morning. With this assurance, we can have the system in operation by Friday afternoon.

Additional Benefits
We are now using NTR PCs at our customer sites. I am very familiar with the setup and installation of these machines. By purchasing a brand of computer currently in use, we will not have to worry about extra time spent learning new installation and operation procedures. In addition, we know that all our software is fully compatible with NTR products.

The warranty on the PC905 is for one year. After the warranty period, the equipment is covered by the service plan we have for other computers and printers.

CONCLUSION
I recommend that we purchase the NTR PC905 to replace our file server. It meets or exceeds all criteria for performance, price, and installation.

EXAMPLE 12 (*continued*)

ABC Format 13: Proposal

Abstract

- Purpose of proposal
- Main need of the readers
- Main feature of what you are offering
- Main benefit to reader of the feature noted above
- Main sections to follow

Body

- Problem or need and its significance
- Proposed solution or approach
- People to be used and their qualifications
- Schedule to be followed
- Cost of what is being proposed

Conclusion

- Restate a main feature or benefit or both
- Make clear your interest in the work
- State what should happen next

Helpful Hints

Long and short proposals cover the same basic topics—needs, features, and benefits. Yet long proposals respond to more complex projects and are presented as formal documents, much like formal reports (see ABC Format 15). Whether long or short, proposals can be either in-house documents or sales proposals directed to customers.

DATE: October 6, 2000
TO: Gary Lane
FROM: Jeff Bilstrom *JB*
SUBJECT: Logo Proposal for Montrose Senior Citizens Center

My job as director of public relations is to get the Montrose name firmly entrenched in the minds of metro Atlanta residents. Having reviewed the contacts we have with the public, I believe we are sending a confusing message about the many services we offer retired citizens.

To remedy the problem, I propose we adopt a logo to serve as an umbrella for all services and agencies supported by the Montrose Senior Citizens Center. This proposal gives details about the problem and the proposed solution, including costs.

The Problem

The lack of a logo presents a number of problems related to marketing the center's services and informing the public. Here are a few:

- The letterhead mentions the organization's name in small type, with none of the impact that an accompanying logo would have.
- The current brochure needs the flair that could be provided by a logo on the cover page, rather than just the page of text and headings we now have.
- Our 14 vehicles are difficult to identify because there is only the lettered organization name on the sides, without any readily identifiable graphic.
- The sign in front of our campus, a main piece of free advertising, could better spread the word about Montrose if it contained a catchy logo.
- Other signs around campus could display the logo, as a way of reinforcing our identity and labeling buildings.

It's clear that without a logo, the Montrose Senior Citizens Center misses an excellent opportunity to educate the public about its services.

The Solution

A professionally designed logo could provide the Montrose Senior Citizens Center with a more distinct identity. Helping to tie together all branches of our operation, it would give the public an easy-to-recognize symbol. As a result, there would be a stronger awareness of the center on the part of users and financial contributors.

EXAMPLE 13 (pp. 98–99)
Proposal

Source: Technical Writing: A Practical Approach, 4th ed. (pp. 361–362) by W. S. Pfeiffer, 2000, Upper Saddle River, NJ: Prentice Hall. Reprinted by permission.

The new logo could be used immediately to do the following:
- Design and print letterhead, envelopes, business cards, and a new brochure.
- Develop a decal for all company vehicles that would identify them as belonging to Montrose.
- Develop new signs for the entire campus, to include a new sign for the entrance to the campus, one sign at the entrance to the Blane Workshop, and one sign at the entrance to the Administration Building.

Cost

Developing a new logo can be quite expensive. However, I have been able to get the name of a well-respected graphic artist willing to donate his services in the creation of a new logo. We need only give him some general guidelines to follow and then choose from among eight to ten rough sketches. Once a decision is made, the artist will provide a camera-ready copy of the new logo.

■ Design charge	$0.00
■ Charge for new letterhead, envelopes, business cards, and brochures (min. order)	545.65
■ Decal for vehicles (14 @ $50.00 + 4%)	728.00
■ Signs for campus	415.28
Total Cost	$1,688.93

Conclusion

As the retirement population of Atlanta increases in the next few years, there will be a much greater need for the services of the Montrose Senior Citizens Center. Because of that need, it's in our best interest to keep this growing market informed about the organization.

I'll stop by later this week to discuss any questions you might have about this proposal.

ABC Format 14: Progress Report

Abstract

- Purpose of report
- Overview of project
- Survey of progress since last report

Body

- Tasks completed since last report—organized by task or time or both
- Clear reference to deadends that have taken time but yielded no results
- Explanation of delays or incomplete work
- Work remaining on project(s)—organized by task or time or both
- Reference to attachments that may contain more specific information

Conclusion

- Brief restatement of work since last reporting period
- Expression of confidence, or concern, about overall work on project
- Indication of willingness to make adjustments the reader may suggest

Helpful Hints

Progress reports are usually organized by task or by time. The reader, perhaps your supervisor or your customer, wants details about the status of a project. Whereas you should put forth the best case for the work you have completed, be certain not to overstate what you have done.

MASTMAN SAFETY RESEARCH CONSULTANTS, Inc.

MEMORANDUM

DATE: June 11, 2000
TO: Kerry Camp, Vice President of Domestic Operations
FROM: Scott Sampson, Manager of Personnel *SS*
SUBJECT: Progress Report on Training Project

INTRODUCTORY SUMMARY

On May 21 you asked that I study ways our firm can improve training for technical employees in all domestic offices. We agreed that the project would take about six or seven weeks and involve three phases:

Phase 1: Make phone inquiries to competing firms
Phase 2: Send a survey to our technical people
Phase 3: Interview a cross section of our technical employees

I have now completed Phase 1 and part of Phase 2. My observation thus far is that the project will offer many new directions to consider for our technical training program.

WORK COMPLETED

In the first week of the project, I had extensive phone conversations with people at three competing firms about their training programs. Then in the second week, I wrote and sent out a training survey to all technical employees in Mastman's domestic offices.

Phone Interviews

I contacted three firms for whom we have done similar favors in the past: Simkins Consultants, Judd & Associates, and ABG Engineering. Here is a summary of my conversations:

1. Simkins Consultants
   Talked with Harry Roland, training director, on May 23. Harry said that his firm has most success with internal training seminars. Each technical person completes several one- or two-day seminars every year. These courses are conducted by in-house experts or external consultants, depending on the specialty.

EXAMPLE 14 (pp. 101–102)

Progress report

Source: Adapted from *Technical Writing: A Practical Approach,* 4th ed. (pp. 280–281) by W. S. Pfeiffer, 2000, Upper Saddle River, NJ: Prentice Hall. Reprinted by permission.

Kerry Camp   Page 2
June 11, 2000

2. Judd & Associates
   Talked with Jan Tyler, manager of engineering, on May 23. Jan said that Judd,
   like Simkins, depends mostly on internal seminars. But Judd spreads these
   seminars over one or two weeks, rather than teaching intensive courses in one
   or two days. Judd also offers short "technical awareness" sessions at the lunch
   hour every two weeks. In-house technical experts give informal presentations
   on some aspect of their research or fieldwork.
3. ABG Engineering
   Talked with Newt Mosely, personnel coordinator, on May 27. According to Newt,
   ABG's training program is much as it was two decades ago. Most technical people at high levels go to one seminar a year, usually sponsored by professional
   societies or local colleges. Other technical people get little training beyond what
   is provided on the job. In-house training has not worked well, mainly because
   of schedule conflicts with engineering jobs.

Internal Survey
   After completing the phone interviews noted, I began the survey phase of the
project. Last week, I finished writing the survey, had it reproduced, and sent it with
a cover memo to all 450 technical employees in domestic offices. The deadline for
returning it to me is June 17.

WORK PLANNED
   With phone interviews finished and the survey mailed, I foresee the following
schedule for completing the project:
June 17:   Surveys returned
June 18–21:   Surveys evaluated
June 24–28:   Trips taken to all domestic offices to interview a cross section of
              technical employees
July 3:   Submission of final project report to you

CONCLUSION
   My interviews with competitors gave me a good feel for what technical training
might be appropriate for our staff. Now I am hoping for a high-percentage return on
the internal survey. That phase will prepare a good foundation for my on-site interviews later this month. I believe this major corporate effort will upgrade our technical training considerably.
   I would be glad to hear any suggestions you may have about my work on the
rest of the project. In particular, please call if you have any specific questions you
want asked during the on-site interviews (ext. 348).

EXAMPLE 14 (*continued*)

ABC Format 15: Formal Report

Abstract

- Cover/title page
- Letter or memo of transmittal
- Table of contents
- List of illustrations
- Executive summary
- Introduction

Body

- Discussion sections
- [Appendices—appear after text but support Body section]

Conclusion

- Conclusions
- Recommendations

Helpful Hints

The term formal report refers to documents of a certain length (at least 6 to 10 pages of text) and degree of formality (usually bound, with formal covers and graphics). They cover complex projects and can include one or more types of information noted in previous ABC models. One part of the formal report—the executive summary—is especially important to decision-makers.

**STUDY OF WILDWOOD CREEK
WINSLOW, GEORGIA**

Prepared for:

The City of Winslow

Prepared by:

Christopher S. Rice, Hydro/Environmental Engineer
Adagio, Inc.

November 28, 2000

EXAMPLE 15 (pp. 104–119)
Formal report
Source: Adapted from *Technical Writing: A Practical Approach,* 4th ed. (pp. 313–328) by W. S. Pfeiffer, 2000, Upper Saddle River, NJ: Prentice Hall. Reprinted by permission.

12 Peachtree Street
Atlanta, GA 30056
(404) 555-7524

Adagio Project #96-119
November 28, 2000

Adopt-a-Stream Program
City of Winslow
300 Lawrence Street
Winslow, Georgia 30000

Attention: Ms. Elaine Sykes, Director

STUDY OF WILDWOOD CREEK
WINSLOW, GEORGIA

We have completed our seven-month project on the pollution study of Wildwood Creek. This project was authorized on May 19, 2000. We performed the study in accordance with our original proposal, No. 14-P72, dated April 24, 2000.

This report mentions all completed tests and discusses the test results. Wildwood Creek scored well on many of the tests, but we are concerned about several problems—such as the level of phosphates in the stream. The few problems we observed during our study have led us to recommend that several additional tests should be conducted.

Thank you for the opportunity to complete this project. We look forward to working with you on further tests for Wildwood Creek and other waterways in Winslow.

Sincerely,

Christopher S. Rice, P.E.

Christopher S. Rice, P.E.
Hydro/Environmental Engineer

*(**continues**)*

TABLE OF CONTENTS

EXAMPLE 15 (*continued*)

LIST OF ILLUSTRATIONS

(*continues*)

EXECUTIVE SUMMARY

The City of Winslow hired Adagio, Inc., to perform a pollution study of Wildwood Creek. The section of the creek that was studied is a one-mile-long area in Burns Nature Park, from Newell College to U.S. Highway 42. The study lasted seven months.

Adagio completed 13 tests on four different test dates. Wildwood scored fairly well on many of the tests, but there were some problem areas. For example, high levels of phosphates were uncovered in the water. The phosphates were derived either from fertilizer or from animal and plant matter and waste. Also uncovered were small amounts of undesirable water organisms that are tolerant to pollutants and can survive in harsh environments.

Adagio recommends that (1) the tests done in this study be conducted two more times, through spring 2001, (2) other environmental tests be conducted, as listed in the conclusions and recommendations section, and (3) a voluntary cleanup of the creek be scheduled. With these steps, we can better analyze the environmental integrity of Wildwood Creek.

1

EXAMPLE 15 (*continued*)

INTRODUCTION

Adagio, Inc., has conducted a follow-up to a study completed in 1990 by Ware County on the health of Wildwood Creek. This introduction describes the project site, scope of our study, and format for this report.

PROJECT DESCRIPTION

By law, all states must clean up their waterways. The State of Georgia shares this responsibility with its counties. Ware County has certain waterways that are threatened and must be cleaned. Wildwood Creek is one of the more endangered waterways. The portion of the creek that was studied for this report is a one-mile stretch in Burns Nature Park between Newell College and U.S. Highway 42.

SCOPE OF STUDY

The purpose of this project was to determine whether the health of the creek has changed since the previous study in 1990. Both physical and chemical tests were completed. The nine physical tests were as follows:

- Air temperature
- Water temperature
- Water flow
- Water appearance
- Habitat description
- Algae appearance
- Algae location
- Visible litter
- Bug count

The four chemical tests were as follows:

- pH
- Dissolved oxygen (DO)
- Turbidity
- Phosphate

REPORT FORMAT

This report includes three main sections:

1. Field Investigation: a complete discussion of all the tests performed
2. Test Comparison: charts of the test results and comparisons
3. Conclusions and Recommendations

2

(*continues*)

FIELD INVESTIGATION

Regulators have repeatedly called attention to environmental violations involving Wildwood Creek. Many factors can generate pollution and affect the overall health of the creek. In 1990, the creek was studied in the context of a study of all water systems in Ware County. Wildwood Creek was determined to be one of the more threatened creeks in the county.

The city needed to learn if much has changed in the past seven years, so Adagio was hired to perform a variety of tests on the creek. Our effort involved a more in-depth study than that done in 1990. Tests were conducted four times over a seven-month period. The 1990 study lasted only one day.

The field investigation included two categories of tests: physical tests and chemical tests.

PHYSICAL TESTS

The physical tests covered a broad range of environmental features. This section will discuss the importance of the tests and some major findings. The Test Comparison section on page 8 includes a table that lists results of the tests and the completion dates. The test types were as follows: air temperature, water temperature, water flow, water appearance, habitat description, algae appearance, algae location, visible litter, and bug count.

Air Temperature

The temperature of the air surrounding the creek will affect life in the water. Unseasonable air temperature will determine if life can grow in or out of the water.

Three of the four tests were performed in the warmer months. Only one was completed on a cool day. The difference in temperature from the warmest to coolest day was 10.5°C, an acceptable range.

Water Temperature

The temperature of the water determines which species will be present. Also affected are the feeding, reproduction, and metabolism of these species. If there are one or two weeks of high temperature, the stream is unsuitable for most species. If water temperature changes more than 1° to 2°C in 24 hours, thermal stress and shock can occur, killing much of the life in the creek.

During our study, the temperature of the water averaged 1°C cooler than the temperature of the air. The water temperature did not get above 23°C or below 13°C. These ranges are acceptable by law.

3

EXAMPLE 15 (*continued*)

Water Flow

The flow of the water will influence the type of life in the stream. Periods of high flow can cause erosion to occur on the banks and sediment to cover the streambed. Low water flow can decrease the living space and deplete the oxygen supply.

The flow of water was at the correct level for the times of year the tests were done—except for June, which had a high rainfall. With continual rain and sudden flash floods, the creek was almost too dangerous for the study to be performed that month.

In fact, in June we witnessed the aftermath of one flash flood. Figure 1 shows the creek with an average flow of water, and Figure 2 shows the creek during the flood. The water's average depth is 10 inches. During the flash flood, the water level rose and fell 10 feet in about one hour. Much dirt and debris were washed into the creek, while some small fish were left on dry land as the water receded.

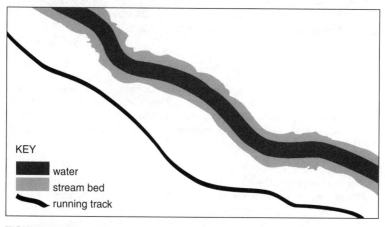

KEY
- water
- stream bed
- running track

FIGURE 1 Wildwood Creek—Normal Water Level

4

(continues)

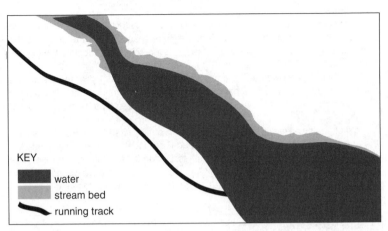

KEY

water

stream bed

running track

FIGURE 2 Wildwood Creek—Flash Flood Water Level

Water Appearance

The color of the water gives a quick but fairly accurate view of the health of the creek. If the water is brown or dirty, then silt or human waste may be present. Black areas of water may contain oil or other chemical products.

On each of the four test days, the water was always clear. Thus the appearance of the creek water was considered excellent.

Habitat Description

The habitat description concerns the appearance of the stream and its surroundings. Important criteria are (1) the number of pools and (2) the number of ripples—that is, points where water flows quickly over rocky areas. Both pools and ripples provide good locations for fish and other creatures to live and breed.

In describing habitat, Adagio also evaluates the amount of sediment at the bottom of the stream. Too much sediment tends to cover up areas where aquatic life lays eggs and hides them from predators. We also evaluate the stability of the stream banks; a stable bank indicates that erosion has not damaged the habitat. Finally, we observe the amount of stream cover. Such vegetation helps keep soil in place on the banks.

EXAMPLE 15 (*continued*)

Wildwood Creek tested fairly well for habitat. The number of pools and ripples was about average for such creeks. Stream deposits and stream bank stability were average to good, and stream cover was good to excellent. For more detail about test results, see the chart in the Test Comparison section on page 8.

Algae Appearance and Location
Algae is naturally present in any creek. The amount of algae can be a warning of pollution in the water. If algae is growing out of control, disproportionate amounts of nutrients such as nitrogen or phosphate could be present. These chemicals could come from fertilizer washed into the creek. Excessive amounts of algae cause the oxygen level to drop when they die and decompose.

During the four studies, algae was everywhere, but it was heaviest on the rocks in the ripples of the creek. The algae was always brown and sometimes hairy.

Visible Litter
Litter can affect the habitat of a creek. While some litter has chemicals that can pollute the water, other litter can cover nesting areas and suffocate small animals. Whether the litter is harmful or not, it is always an eyesore.

On all four test dates, the litter we saw was heavy and ranged from tires to plastic bags. Some of the same trash that was at the site on the first visit was still there seven months later.

Bug Count
The bug count is a procedure that begins by washing dirt and water onto a screen. As water drains, the dirt with organisms is left on the screen. The bugs are removed and classified. Generally, the lower the bug count, the higher the pollution levels. Bug counts were considered low to average.

Two types of aquatic worms were discovered every time during our count, but in relatively small numbers. In addition, the worms we observed are very tolerant of pollution and can live in most conditions. Finally, we observed only two crayfish, animals that are somewhat sensitive to pollution.

CHEMICAL TESTS
While physical tests cover areas seen with the naked eye, chemical tests can uncover pollutants that are not so recognizable. Certain chemicals can wipe out all life in a creek. Other chemicals can cause an overabundance of one life-form, which in turn could kill more sensitive animals.

A chart of results of chemical tests is included in the Test Comparison section on page 8. The chemical tests that Adagio performed were pH, dissolved oxygen (DO), turbidity, and phosphate.

6

(continues)

pH

The pH test is a measure of active hydrogen ions in a sample. The range of the pH test is 0–14. If the sample is in the range of 0–7, it is acidic; but if the sample is in the range of 7.0–14, it is basic. By law, the pH of a water sample must be within the range of 6.0 to 8.5.

For the tests we completed, the water sample was always 7.0, which is very good for a creek.

Dissolved Oxygen (DO)

Normally, oxygen dissolves readily into water from surface air. Once dissolved, it diffuses slowly in the water and is distributed throughout the creek. The amount of DO depends on different circumstances. Oxygen is always highest in choppy water, just after noon, and in cooler temperatures.

In many streams, the level of DO can become critically low during the summer months. When the temperature is warm, organisms are highly active and consume the oxygen supply. If the amount of DO drops below 3.0 ppm (parts per million), the area can become stressful for the organisms. An amount of oxygen that is 2.0 ppm or below will not support fish. DO that is 5.0 ppm to 6.0 ppm is usually required for growth and activity of organisms in the water.

According to the Water Quality Criteria for Georgia, average daily amounts of DO should be 5.0 ppm with a minimum of 4.0 ppm. Wildwood Creek scored well on this test. The average amount of DO in the water was 6.9 ppm, with the highest amount being 9.0 ppm on November 19, 2000.

Turbidity

Turbidity is the discoloration of water caused by sediment, microscopic organisms, and other matter. One major factor of turbidity is the level of rainfall before a test. Three of our tests were performed on clear days with little rainfall. On these dates, the turbidity of Wildwood was always 1.0, the best that creek water can score. The fourth test, which scored worse, occurred during a rainy period.

Phosphate

Phosphorus occurs naturally as phosphates—for example, orthophosphates and organically bound phosphates. Orthophosphates are phosphates that are formed in fertilizer, while organically bound phosphates can form in plant and animal matter and waste.

Phosphate levels higher than 0.03 ppm contribute to increased plant growth. If phosphate levels are above 0.1 ppm, plants can grow out of control. The phosphate level of Wildwood was always 0.5 ppm, much higher than is desirable.

EXAMPLE 15 (*continued*)

TEST COMPARISON

There was little change from each of the four test dates. The only tests that varied greatly from one test to another were air temperature, water temperature, water flow, and DO. On the basis of these results, it would appear that Wildwood Creek is a relatively stable environment.

TABLE 1 Physical Tests

TEST	5/26/00	6/25/00	9/24/00	11/19/00
Air Temperature in °C	21.5	23.0	24.0	13.5
Water Temperature in °C	20.0	22.0	23.0	13.0
Water Flow	Normal	High	Normal	Normal
Water Appearance	Clear	Clear	Clear	Clear
Habitat Description				
Number of Pools	2.0	3.0	2.0	5.0
Number of Ripples	1.0	2.0	2.0	2.0
Amount of Sediment Deposit	Average	Average	Good	Average
Stream Bank Stability	Average	Good	Good	Good
Stream Cover	Excellent	Good	Excellent	Good
Algae Appearance	Brown	Brown/hairy	Brown	Brown
Algae Location	Everywhere	Everywhere	Attached	Everywhere
Visible Litter	Heavy	Heavy	Heavy	Heavy
Bug Count	Low	Average	Low	Average

TABLE 2 Chemical Tests

TEST	5/26/00	6/25/00	9/24/00	11/19/00
pH	7.0	7.0	7.0	7.0
Dissolved Oxygen (DO)	6.8	6.0	5.6	9.0
Turbidity	1.0	3.0	1.0	1.0
Phosphate	.50	.50	.50	.50

(*continues*)

CONCLUSIONS AND RECOMMENDATIONS

This section includes the major conclusions and recommendations from our study of Wildwood Creek.

CONCLUSIONS

Generally, we were pleased with the health of the stream bank and its floodplain. The area studied has large amounts of vegetation along the stream, and the banks seem to be sturdy. The floodplain has been turned into a park, which handles floods in a natural way. Floodwater in this area comes in contact with vegetation and some dirt. Floodwater also drains quickly, which keeps sediment from building up in the creek.

However, we are concerned with the number and types of animals uncovered in our bug counts. Only two bug types were discovered, and these were types quite tolerant to pollutants. The time of year these tests were performed could affect the discovery of some animals. However, the low count still should be considered a possible warning sign about water quality. Phosphate levels were also high and probably are the cause of the large amount of algae.

We believe something in the water is keeping sensitive animals from developing. One factor that affects the number of animals discovered is the pollutant problems in the past (see Appendix A). The creek may still be in a redevelopment stage, thus explaining the small numbers of animals.

RECOMMENDATIONS

On the basis of our conclusions, we recommend the following actions for Wildwood Creek:

1. Conduct the current tests two more times, through spring 2001. Spring is the time of year that most aquatic insects are hatched. If sensitive organisms are found then, the health of the creek could be considered to have improved.
2. Add testing for nitrogen. With the phosphate level being so high, nitrogen might also be present. If it is, then fertilizer could be in the water.
3. Add testing for human waste. Some contamination may still be occurring.
4. Add testing for metals, such as mercury, that can pollute the water.
5. Add testing for runoff water from drainage pipes that flow into the creek.
6. Schedule a volunteer cleanup of the creek.

With a full year of study and additional tests, the problems of Wildwood Creek can be better understood.

9

EXAMPLE 15 (*continued*)

APPENDIX A
Background on Wildwood Creek

Wildwood Creek begins from tributaries on the northeast side of the city of Winslow. From this point, the creek flows southwest to the Chattahoochee River. Winslow Wastewater Treatment Plant has severely polluted the creek in the past with discharge of wastewater directly into the creek. Wildwood became so contaminated that signs warning of excessive pollution were posted along the creek to alert the public.

Today, all known wastewater discharge has been removed. The stream's condition has dramatically improved, but nonpoint contamination sources continue to lower the creek's water quality. Nonpoint contamination includes sewer breaks, chemical dumping, and storm sewers.

Another problem for Wildwood Creek is siltration. Rainfall combines with bank erosion and habitat destruction to wash excess dirt into the creek. This harsh action destroys most of the macroinvertebrates. At the present time, Wildwood Creek may be one of the more threatened creeks in Ware County.

(*continues*)

APPENDIX B
Water Quality Criteria for Georgia

All waterways in Georgia are classified in one of the following categories: fishing, recreation, drinking, and wild and scenic. Different protection levels apply to the different uses. For example, the protection level for dissolved oxygen is stricter in drinking water than fishing water. All water is supposed to be free from all types of waste and sewage that can settle and form sludge deposits.

In Ware County, all waterways are classified as "fishing," according to Chapter 391-3-6.03 of "Water Use Classifications and Water Quality Standards" in the Georgia Department of Natural Resources *Rules and Regulations for Water Quality Control.* The only exception is the Chattahoochee River, which is classified as "drinking water supply" and "recreational."

11

EXAMPLE 15 (*continued*)

APPENDIX C

Map 6
Location of City of Winslow Parks and Recreation Facilities

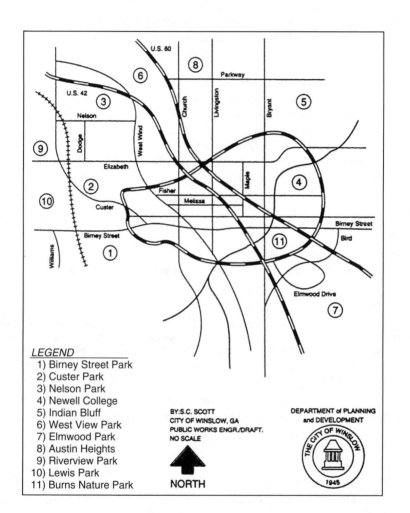

LEGEND
1) Birney Street Park
2) Custer Park
3) Nelson Park
4) Newell College
5) Indian Bluff
6) West View Park
7) Elmwood Park
8) Austin Heights
9) Riverview Park
10) Lewis Park
11) Burns Nature Park

BY:S.C. SCOTT
CITY OF WINSLOW, GA
PUBLIC WORKS ENGR./DRAFT.
NO SCALE

NORTH

DEPARTMENT of PLANNING
and DEVELOPMENT

THE CITY OF WINSLOW
1945

ABC Format 16: Job Letter and Resume
Abstract

- Apply for a specific job
- Refer to ad, mutual friend, or other source of information about the job
- (Optional) Briefly state how you can meet the main need of your potential employer

Body

- Specify your understanding of the reader's main needs
- Provide main qualifications that satisfy these needs (but only highlight points from resume—do NOT simply repeat all resume information)
- Avoid mentioning weak points or deficiencies
- Keep body paragraphs to six or fewer lines
- Use a bulleted or numbered list if it helps draw attention to three or four main points
- Maintain the "you" attitude throughout

Conclusion

- Tie the letter together with one main theme or selling point, as you would a sales letter
- Refer to your resume
- Explain how and when the reader can contact you for an interview

Helpful Hints

In the first examples of job letters and resumes that follow, note that one resume is in what is called "chronological" format and the other is in "functional" format. Also included is a third resume that combines the two formats. All three reflect a clear presentation of basic information about education, work experience, skills, and interests. However, they differ in what they emphasize, as explained below:

- James Sistrunk's resume is considered "chronological." Because he has relevant full-time job experience, he gives it priority—with the most recent work listed first.
- Denise Sanborn's resume is considered "functional." Because she has limited work experience, she wants to focus instead on the "functional" skills acquired in her part-time and summer jobs.
- Karen Patel's combined format resume lists two main skill sets as her main categories, under which particular full- and part-time jobs and their time periods are included.

Use the resume format that best matches your background. Above all, keep your resume looking clean. Remember that its purpose is only to get you to the next stage of the job process, the interview.

1523 River Lane
Worthville, OH 43804
August 6, 2000

Mr. Willard Yancy
Director, Automotive Systems
XYZ Motor Company, Product Development Division
Charlotte, NC 28202

Dear Mr. Yancy:

Recently I have been researching the leading national companies in automotive computer systems. Your job ad in the July 6 *National Business Employment Weekly* caught my eye because of XYZ's innovations in computer-controlled safety systems. I would like to apply for the automotive computer engineer job.

Your advertisement notes that experience in computer systems for machinery or robotic systems would be a plus. I have had extensive experience in the military with computer systems, ranging from a digital communications computer to an air traffic control training simulator. In addition, my college experience includes courses in computer engineering that have broadened my experience. I am eager to apply what I have learned to your company.

My mechanical knowledge was gained from growing up on my family's dairy farm. After watching and learning from my father, I learned to repair internal combustion engines, diesel engines, and hydraulic systems. Then for five years I managed the entire dairy operation.

With my training and hands-on experience, I believe I can contribute to your company. Please contact me at 614/882-2731 if you wish to arrange an interview.

Sincerely,

James M. Sistrunk

James M. Sistrunk

Enclosure: Resume

EXAMPLE 16 (pp. 121–125)
Job Letter and Resume
Source: Technical Writing: A Practical Approach, 4th ed. (pp. 562–565, 569) by W. S. Pfeiffer, 2000, Upper Saddle River, NJ: Prentice Hall. Reprinted by permission.

James M. Sistrunk
1523 River Lane
Worthville, OH 43804
(614) 882-2731

Professional Objective:
To contribute to the research, design, and development of automotive computer control systems

Education:
B.S., Computer Engineering, 1996-present
Columbus College, Columbus, Ohio
Major concentration in Control Systems with minor in Industrial Engineering. Courses included Microcomputer Systems, Digital Control Systems, and several different programming courses.
Computer Repair Technician Certification Training, 1993-1994
U.S. Air Force Technical Training Center, Keesler Air Force Base, Biloxi, Mississippi.
General Computer Systems Option with emphasis on mainframe computers. Student leader in charge of processing and orientation for new students from basic training.

Career Development:
Computer Repair Technician, U.S. Air Force, 1993-1996
Secret Clearance

Responsibilities and duties included:
- Repair of computer systems
- Preventative maintenance inspections
- Diagnostics and troubleshooting of equipment

Accomplishments included:
- "Excellent" score during skills evaluation
- Award of an Air Force Specialty Code "5" skill level

Assistant Manager, Spring Farm, Wootan, Ohio, 1987-1992
Responsible for dairy operations on this 500-acre farm. Developed the management and technical skills; learned to repair sophisticated farm equipment.

Special Skills:
Macintosh desk-top publishing
Windows 95/98
Assembly Language
C++ Programming

References:
Available upon request

EXAMPLE 16 (*continued*)

456 Cantor Way, #245
Gallop, Minnesota 55002
September 3, 2000

Ms. Judith R. Gonzalez
American Hospital Systems
3023 Center Avenue
Randolf, Minnesota 55440

Dear Ms. Gonzalez:

My placement center recently informed me about the Management Trainee opening with Mercy Hospital. As a business major with experience working in hospitals, I wish to apply for the position.

Your job advertisement notes that you seek candidates with a broad academic background in business and an interest in hospital management. At Central State College, I've taken extensive coursework in three major areas in business: finance, marketing, and personnel management. This broad-based academic curriculum has provided a solid foundation for a wide variety of management tasks at Mercy Hospital.

My summer and part-time employment also matches the needs of your position. While attending Central State, I've worked part-time and summers as an assistant in the Business Office at Grady Hospital. That experience has acquainted me with the basics of business management within the context of a mid-sized hospital, much like Mercy.

The enclosed resume highlights the skills that match your Management Trainee opening. I would like the opportunity to talk with you in person and can be reached at 612-111-1111 for an interview.

Sincerely,

Denise Ware Sanborn

Denise Ware Sanborn

(*continues*)

Denise Ware Sanborn
456 Cantor Way, #245
Gallop, Minnesota 55002
612-111-1111

Objective
Entry-level management position in the health care industry. Seek position that includes exposure to a wide variety of management and business-related tasks.

Education
Bachelor of Arts Degree, June 2000
Central State College
Gallop, Minnesota

Major: Business Administration
Grade Point Average: 3.26 of possible 4.0, with 3.56 in all major courses
All college expenses financed by part-time and summer work at Grady Hospital in St. Paul, Minnesota

Skills and Experience
   Finance
        Helped with research for three fiscal year budgets
        Developed new spreadsheet for monthly budget reports
        Wrote accounts payable correspondence
   Marketing
        Solicited copy from managers for new brochure
        Designed and edited new brochure
        Participated in team visits to ten area physicians
   Personnel
        Designed new performance appraisal form for secretarial staff
        Interviewed applicants for Maintenance Department jobs
        Coordinated annual training program for nursing staff

Awards
2000 Arden Award for best senior project in the Business Administration Department (paper that examined Total Quality Management)

Dean's list for six semesters

References
Academic and work references available upon request

EXAMPLE 16 (*continued*)

Karen S. Patel
300 Park Drive
Burtingdale, New York 20092

Home: (210) 400-2112　　　　　　　　　　　　　**Messages:** (210) 400-0111

OBJECTIVE　　　　Position as in-house technical writer and as trainer in communication skills

EDUCATION　　　　**Sumpter College, Marist, Vermont**
M.S. in Technical Communication, GPA: 4.0
December 1997

Warren College, Aurora, New York
M.A. in English, Cum Laude, June 1994

University of Bombay, India
B.A. in English, First Class Honors, June 1991

EMPLOYMENT
Editing/
Writing
Public Relations Office, Sumpter College, 1997-present
Administrative Assistant: Write press releases and conduct interviews. Publish news stories in local newspapers and in *Sumpter Express*. Edit daily campus newsletter.

Hawk Newspapers, Albany, New York, 1992-1993
Warren College Internship: Covered and reported special events; conducted interviews; assisted with proofreading, layout, headline count. Scanned newspapers for current events; conducted research for stories. Published feature stories.

Teaching/
Research
Sumpter College, Marist, Vermont, 1996-1997
Teaching Assistant: Tutored English at the Writing Center, answered "Grammar Hotline" phone questions, edited and critiqued student papers, taught English to non-English speakers, and helped students prepare for Regents exams.

Warren College, Aurora, New York, 1993-1994
Teaching Assistant: Taught business writing, supervised peer editing and in-class discussions, held student conferences, and graded student papers.

Research Assistant: Verified material by checking facts, wrote brief reports related to research, researched information and bibliographies.

COMPUTER SKILLS　　　Wordperfect, Microsoft Word, Pagemaker, Unix, Excel

REFERENCES　　　Available upon request

Appendix B: Writing Handbook

*C*hapter 1 divided the writing process into three stages: planning, drafting, and revising. This appendix will help you fix problems that occur during the revision stage of writing. It includes alphabetized entries on style, grammar, and usage. These three terms are defined as follows:

Style: features such as word choice, sentence length, paragraph organization, and active and passive voice. These are matters of choice, not right and wrong. For example, both the wordy and concise versions of a passage may be grammatically correct, but the concise version is a better stylistic choice.

Grammar: actual rules that determine how language should be used, as opposed to stylistic options one can choose. Examples of such "right/wrong" issues include comma placement and correct use of modifiers.

Usage: contexts for the correct use of problem words—especially word pairs such as "effect/affect" and "complement/compliment."

Each entry in this appendix has two parts: an explanation of the meaning and one or more correct examples.

Of course, when you're preparing a document for your boss or client, you don't separate problems by style, grammar, or usage. You just want to produce a draft that is easy to read and free of errors. Thus all entries in this appendix are alphabetized for easy reference. Following are two separate tables of contents—alphabetical and topical:

ALPHABETIZED TABLE OF CONTENTS

TOPICAL TABLE OF CONTENTS

■ *STYLE*

▇ *GRAMMAR*

▇ *USAGE*

A/AN

These two words are different forms of the same article. *A* occurs before words that start with consonants or consonant sounds. Examples:

- *a* three-pronged plug
- *a* once-in-a-lifetime job (though its first letter is a vowel, *once* begins with the consonant sound of "w")
- *a* historic moment (because *history* begins with a consonant sound)

An occurs before words that begin with vowels or vowel sounds. Examples:

- *an* earthly paradise
- *an* hour (though its first letter is a consonant, *hour* begins with a vowel sound)
- *an* eager new employee

A LOT/ALOT

The correct form is the two-word phrase *a lot*. (*Alot* is not acceptable usage.) Though acceptable in casual discourse, *a lot* should be replaced by more formal diction in technical writing. Examples:

- My father had *a lot* of patience.
- They retrieved *many* [not *a lot of*] soil samples from the site.

ABBREVIATIONS: RULES

Technical writing uses many abbreviations. Without this shorthand form, you would produce much longer reports and proposals—without adding content. Follow six basic rules in your use of abbreviations:

■ *DO NOT USE ABBREVIATIONS WHEN CONFUSION MAY RESULT*

When you want to use a term just once or twice *and* you are not certain readers will understand an abbreviation, write out the term rather than abbreviating it. Example:

> They were required to remove creosote from the site, according to the directive from the Environmental Protection Agency.

Even though *EPA* is the accepted abbreviation for the agency mentioned, you should write out the name in full *if* you are using the term only once to an audience that may not understand it.

■ USE PARENTHESES FOR CLARITY

When you use a term *more* than twice and are not certain your readers will understand it, write out the term the first time it is used and place the abbreviation in parentheses. Then use the abbreviation in the rest of the document. In long reports or proposals, however, you may need to repeat the term with parenthetical abbreviation in key places. Example:

> According to the directive from the Environmental Protection Agency (EPA), the builders were required to remove the creosote from the construction site. Furthermore, the directive noted that builders could expect to be visited by EPA inspectors every other week.

■ INCLUDE A GLOSSARY WHEN THERE ARE MANY ABBREVIATIONS

When a document contains many abbreviations that may not be understood, include a glossary at the beginning or end of the document. A glossary simply collects all the terms and abbreviations and places them in one location, for easy reference.

■ USE ABBREVIATIONS FOR UNITS OF MEASURE

Most technical documents use abbreviations for units of measure. Do not include a period unless the abbreviation could be confused with a word. Examples:

> mi, ft, oz, in., gal., lb

Note that units-of-measurement abbreviations have the same form for both singular and plural amounts. Examples:

> $1/2$ in., 1 in., 5 in.

■ AVOID SPACING AND PERIODS

Avoid internal spacing and internal periods in most abbreviations that contain all capital letters. Examples:

> ASTM, EPA, ASEE

Some common exceptions include professional titles and degrees such as P.E., B.S., and B.A.

■ *Be Careful with Company Names*

Abbreviate a company or other organizational name only when you are certain that officials from the organization consider the abbreviation appropriate. IBM (for the company) and UCLA (for the university) are examples of commonly accepted organizational abbreviations. When in doubt, follow the preceding rule regarding parentheses—that is, write the name in full the first time it is used, followed by the abbreviation in parentheses.

ABBREVIATIONS: COMMON EXAMPLES

The following common abbreviations are appropriate for most technical writing. They are placed into three main categories of measurements, locations, and titles:

■ *Measurements*

Use these abbreviations only when you place numbers before a measurement.

ac: alternating current	f: farad
amp: ampere	F: Fahrenheit
bbl: barrel	Fbm: foot board measure
Btu: British thermal unit	fig.: figure
bu: bushel	fl oz: fluid ounce
C: Celsius	ft: foot (feet)
cal: calorie	g: gram
cc: cubic centimeter	gal. or gl: gallon
circ: circumference	gpm: gallons per minute
cm: centimeter	hp: horsepower
cos: cosine	hr: hour
cot: cotangent	Hz: hertz
cps: cycles per second	in.: inch
cu ft: cubic foot (feet)	j: joule
db: decibel	K: Kelvin
dc: direct current	ke: kinetic energy
dm: decimeter	kg: kilogram
doz or dz: dozen	km: kilometer

kw: kilowatt

kwh: kilowatt hour

l: liter

lb: pound

lin: linear

lm: lumen

log: logarithm

m: meter

mm: millimeter

min: minute

oz: ounce

ppm: parts per million

psf: pounds per square foot

psi: pounds per square inch

pt: pint

qt: quart

rev: revolution

rpm: revolutions per minute

sec: second

sq: square

sq ft: square foot (feet)

T: ton

tan: tangent

v: volt

va: volt-ampere

w: watt

wk: week

wl: wavelength

yd: yard

yr: year

■ *LOCATIONS*

Use these common abbreviations for addresses—on envelopes and letters, for example. However, write out the words in full in other contexts.

AL: Alabama

AK: Alaska

AS: American Samoa

AZ: Arizona

AR: Arkansas

CA: California

CZ: Canal Zone

CO: Colorado

CT: Connecticut

DE: Delaware

DC: District of Columbia

FL: Florida

GA: Georgia

GU: Guam

HI: Hawaii

ID: Idaho

IL: Illinois

IN: Indiana

IA: Iowa

KS: Kansas

KY: Kentucky

LA: Louisiana

ME: Maine

MD: Maryland

MA: Massachusetts

MI: Michigan

MN: Minnesota

MS: Mississippi

MO: Missouri

MT: Montana

NE: Nebraska	VT: Vermont
NV: Nevada	VI: Virgin Islands
NH: New Hampshire	VA: Virginia
NJ: New Jersey	WA: Washington
NM: New Mexico	WV: West Virginia
NY: New York	WI: Wisconsin
NC: North Carolina	WY: Wyoming
ND: North Dakota	AB: Alberta
OH: Ohio	BC: British Columbia
OK: Oklahoma	MB: Manitoba
OR: Oregon	NB: New Brunswick
PA: Pennsylvania	NF: Newfoundland
PR: Puerto Rico	NWT: North West Territories
RI: Rhode Island	NS: Nova Scotia
SC: South Carolina	ON: Ontario
SD: South Dakota	PEI: Prince Edward Island
TN: Tennessee	QC: Quebec
TX: Texas	SK: Saskatchewan
UT: Utah	YK: Yukon

■ *TITLES*

Some of these abbreviations go before the name (such as Dr., Mr., and Messrs.), while others go after the name (such as college degrees, Jr., and Sr.).

Atty.: Attorney

B.A.: Bachelor of Arts

B.S.: Bachelor of Science

D.D.: Doctor of Divinity

Dr.: Doctor (used mainly with medical and dental degrees but also with other doctorates)

Drs.: plural of Dr.

D.V.M.: Doctor of Veterinary Medicine

Hon.: Honorable

Jr.: Junior

LL.D.: Doctor of Laws

M.A.: Master of Arts

M.S.: Master of Science

M.D.: Doctor of Medicine

Messrs.: Plural of Mr.

Mr.: Mister

Mrs.: used to designate a married, widowed, or divorced woman

Ms.: used increasingly for all women, especially when one is uncertain about a woman's marital status

Ph.D.: Doctor of Philosophy

Sr.: Senior

ACCEPT/EXCEPT

These words have different meanings and often are different parts of speech. *Accept* is a verb that means "to receive." *Except* is a preposition or verb and means "to make an exception or special case of." Examples:

- I *accepted* the service award from my office manager.
- Everyone *except* Jonah attended the marine science lecture.
- The company president *excepted* me from the meeting because I had an important sales call to make the same day.

ACCURATE WORDING

Good technical writing style demands accuracy in phrasing. Indeed, many technical professionals place their reputation and even their financial lives on the line with every document that goes out their door. That fact makes clear the importance of taking your time on any editing pass that deals with accuracy of phrasing. Accuracy often demands more words, not fewer. The main rule, then, is never to sacrifice clarity for conciseness. Following are some rules for writing accurately. They also will help you to avoid unnecessary exposure to liability.

■ DISTINGUISH FACTS FROM OPINIONS

Always identify your opinions and judgments as such by using phrases like "we recommend," "we believe," "we suggest," and "in our opinion." If you want to avoid repetitious use of these phrases, group your opinions into a series or list. Thus a single lead-in can indicate to the reader that opinions, not facts, are forthcoming. Examples:

- In our opinion, spread footings would be an acceptable foundation for the building you plan at the site.
- On the basis of our site visit and our experience at similar sites, we believe that (1) ...., (2) ...., and (3) ....

■ *INCLUDE OBVIOUS QUALIFYING STATEMENTS WHEN NEEDED*

Be aware of possible misinterpretations by those who may not know that even technical fields can be inexact. Write carefully, but without becoming overly defensive. In other words, your readers—especially those who are nontechnical—want to know what you did do *and* what you did not do. Example:

> Our summary of soil conditions is based only on information obtained during a brief visit to the site. We did not drill any soil borings.

■ *AVOID ABSOLUTES UNLESS YOU MEAN THEM*

Some words may convey a stronger meaning than you intend. One notable example is *minimize,* which means to reduce to the lowest possible level or amount. If a report claims that a particular piece of equipment will "minimize" breakdowns on the assembly line, the passage could be interpreted as an absolute commitment. In theory, the reader could consider any breakdown at all to be a violation of the report's implications. If instead the writer had used the verb *limit* or *reduce,* the wording would have been more accurate and less open to misunderstanding.

> **Original:** If you follow our recommendations, pollution will be minimized at the site.

> **Revision:** If you follow our recommendations, pollution will be greatly reduced at the site.

The original version would be correct only if the writers were able to guarantee that their recommendations would lead to the lowest possible level of pollution.

ACTIVE AND PASSIVE VOICE

This entry defines the active and passive voices and gives examples of each. It also lists some guidelines for using both voices.

■ *The Meaning of Active and Passive*

Active-voice sentences emphasize the person (or thing) performing the action—that is, somebody (or something) does something ("Mat completed the field study yesterday"). Passive-voice sentences emphasize the action itself—that is, something is being done by somebody ("The field study was completed [by Mat] yesterday"). Here are some other examples of the same thoughts expressed in first the active and then the passive voice:

Examples: Active Voice
1. We *reviewed* aerial photographs in our initial assessment of possible fault activity at the site.
2. The study *revealed* that three underground storage tanks had leaked unleaded gasoline into the soil.
3. We *recommend* that you use a minimum concrete thickness of 6 in. for residential subdivision streets.

Examples: Passive Voice
1. Aerial photographs *were reviewed* [by us] in our initial assessment of possible fault activity at the site.
2. The fact that three underground storage tanks had been leaking unleaded gasoline into the soil *was revealed* in the study.
3. *It is recommended* that you use a minimum concrete thickness of 6 in. for residential concrete streets.

Note that passive constructions are wordier than active ones. They tend to leave out the person or thing doing the action. Although occasionally this impersonal approach is appropriate, often the reader becomes frustrated by writing that does not indicate who or what is doing something.

■ *Using Active and Passive Voice*

The fact is that both the active and passive voice have a role in writing. Knowing when to use each is the key. Here are a few guidelines that will help:

Use the Active Voice When You Want to:
1. Emphasize who is responsible for an action ("*We recommend* that you consider our firm for the work.")
2. Stress the name of a company, whether yours or the reader's ("*PineBluff Contracting has expressed* interest in receiving bids for the Baytown project.")
3. Rewrite a top-heavy sentence so that the main idea is up front ("*Figure 1 shows* the approximate locations of the cars that derailed from the northbound freight train.")

4. Pare down the verbiage in your writing, because the active voice is always a shorter construction (see the active and passive examples in previous subsections)

Use the Passive Voice When You Want to:

1. Emphasize the action rather than the person performing it (*"Samples will be sent* directly from the site to our laboratory in Sacramento.")

2. Avoid the kind of egocentric tone that results from repetitious use of "I," "we," and the name of your company (*"The project will be directed* by two programmers from our Boston office.")

3. Break the monotony of writing that relies too heavily on active-voice sentences.

Technical and business writing depends far too much on the passive voice. This stylistic error results in part from the common misperception that passive writing is more "objective." In fact, overuse of passives only makes writing more tedious to read. In modern business and technical writing, most readers prefer the active voice.

ADVICE/ADVISE/INFORM

Advice is a noun that means "suggestion or recommendation." *Advise* is a verb that means "to suggest or recommend." Do not use the verb *advise* as a substitute for *inform,* which means simply "to provide information." Examples:

- The consultant gave us *advice* on starting a new retirement plan for our employees.
- She *advised* us that a 401(k) plan would be useful for all our employees.
- She *informed* [not *advised*] her clients that they would receive her final report by March 15.

AFFECT/EFFECT

These two words cause much confusion. The key to using them correctly is remembering two simple sentences that cover most (but, unfortunately, not all) usage:

1. *Affect* with an "a" is a verb meaning "to influence."

2. *Effect* with an "e" is a noun meaning "result."

Here is one main exception you also need to know: In special instances, *effect* can be a verb that means "to bring about," as in "He effected considerable change when he became a manager." Examples:

- His progressive leadership greatly *affected* the company's future.
- One *effect* of securing the large government contract was the hiring of several more accountants.
- The president's belief in the future of microcomputers *effected* change in the company's approach to office management. (Less wordy alternative: substitute *changed* for *effected change in.*)

AGREE TO/AGREE WITH

Agree to means that you have *consented to* an arrangement, offer, proposal, etc. *Agree with* only suggests that you are *in harmony with* a certain statement, idea, person, etc. Examples:

- Representatives from our firm *agreed to* alter the contract to reflect the new scope of work.
- We *agree with* you that more study may be needed before the plant is built.

ALL RIGHT/ALRIGHT

All right is an acceptable spelling; *alright* is not. *All right* is an adjective that means "acceptable," an exclamation that means "outstanding," or a phrase that means "correct." Examples:

- Sharon suggested that the advertising copy was *all right* for now but that she would want changes next month.
- Upon seeing his article in print, Zach exclaimed *"All right!"*
- The five classmates were *all right* in their response to the trick questions on the quiz.

ALL TOGETHER/ALTOGETHER

All together is used when items or people are being considered in a group, are working in concert, or are in the same location. *Altogether* is a synonym for "entirely" or "completely." Examples:

- *All together,* the sales team included 50 employees from 15 offices.
- The three firms were *all together* in their support of the agency's plan.
- The managers were *all together* in Cleveland for the conference.
- There were *altogether* too many pedestrians walking near the dangerous intersection.

ALLUSION/ILLUSION/DELUSION/ELUSION

These similar sounding words have distinct meanings. Here is a summary of the differences:

1. **Allusion:** a noun meaning "reference," as in your making an allusion to your vacation in a speech. The related verb is *allude.*
2. **Illusion:** a noun meaning "misunderstanding or false perception." It can be physical (as in seeing a mirage) or mental (as in having the false impression that your hair is not thinning when it is).
3. **Delusion:** a noun meaning a belief based on self-deception. Unlike *illusion,* the word conveys a much stronger sense that someone is out of touch with reality, as in having "delusions of grandeur." The related verb is *delude.*
4. **Elusion:** a noun meaning "the act of escaping or avoiding." The more common form is the verb, *elude,* meaning "to escape or avoid."

Examples:

- His report included an *allusion* to the upcoming visit by the government agency in charge of accreditation.
- She harbored an *illusion* that she was certain to receive the promotion. In fact, her supervisor preferred another department member with more experience.
- He had *delusions* that he soon would become company president, even though he started just last week in the mailroom.
- The main point of the report *eluded* him because there was no executive summary.

ALREADY/ALL READY

All ready is a phrase that means "everyone is prepared," whereas *already* is an adverb that means something is finished or completed. Examples:

- They were *all ready* for the presentation to the client.
- George had *already* arrived at the office before the rest of his proposal team members had even left their homes.

ALTERNATELY/ALTERNATIVELY

Because many readers are aware of the distinction between these two words, any misuse can cause embarrassment or even misunderstanding. Follow these guidelines for correct use.

■ *ALTERNATELY*

As a derivative of *alternate, alternately* is best reserved for events or actions that occur "in turns." Example:

> They used a backhoe during some portions of the project. *Alternately,* they switched to using hand shovels.

■ *ALTERNATIVELY*

A derivative of *alternative, alternatively* should be used when two or more choices are being considered. Example:

> We suggest that you use deep foundations at the site. *Alternatively,* you could consider spread footings.

ALUMNA/ALUMNAE/ALUMNUS/ALUMNI

These terms are derived directly from Latin. An *alumna* is a female graduate, whereas an *alumnus* is a male graduate. The plural forms are *alumnae* for females and *alumni* for males. Some writers prefer to use the plural form *alumni* in situations that involve both male and female graduates. However, the different masculine and feminine forms still are maintained in formal English. Examples:

- As a recent *alumna,* she often helped solicit donations from other *alumnae* of her college.
- He was considered a loyal *alumnus* of the college. Along with 100 other *alumni,* he volunteered every month at the Big Brothers Association of downtown Atlanta.

AMOUNT/NUMBER

Amount is used in reference to items that *cannot* be counted, whereas *number* is used to indicate items that *can* be counted. Examples:

- In the last year we have greatly increased the *amount* of computer paper ordered for the Boston office.
- The last year has seen a huge increase in the *number* [not *amount*] of boxes of computer paper ordered for the Boston office.

AND/OR

This awkward expression probably has its origins in legal writing. It means that there are three separate options to be considered: the item before *and/or,* the item after *and/or,* or both items.

Avoid *and/or* because readers may find it confusing, visually awkward, or both. Instead, replace it with the structure used in the previous sentence. That is, write "A, B, or both," *not* "A and/or B." Example:

> The management trainee was permitted to select two seminars from the areas of computer hardware, communication skills, or both [not *computer hardware and/or communication skills*].

ANTICIPATE/EXPECT

These two words are *not* synonyms. Their meanings are distinctly different. *Anticipate* is used when you mean to suggest or state that steps have been taken *beforehand* to prepare for a situation. *Expect* only means you consider something likely to occur. Examples:

- *Anticipating* that the contract will be successfully negotiated, Jones Engineering is hiring three new hydrologists.
- We *expect* [not *anticipate*] that you will encounter semi-cohesive and cohesive soils in your excavations at the Park Avenue site.

APOSTROPHES

Apostrophes can be used for contractions, for some plurals, and for possessives. Only the latter two uses cause confusion. Use an apostrophe to indicate the plural form of a word as a word. Example:

> That redundant paragraph contained seven *area*'s and three *factor*'s in only five sentences.

Although some writers also use apostrophes to form the plurals of numbers and full-cap abbreviations, the current tendency is to include only the "s." Examples:

7s, ABCs, PCBs, P.E.s.

As for possessives, grammar rules seem to vary, depending on the reference book you're reading. Here are some simple guidelines that reflect this author's view of the best current usage:

■ *Multi-Syllabic Nouns Ending in "s"*

Form the possessive of multi-syllabic nouns that end in "s" by adding just an apostrophe, whether the nouns are singular or plural. Examples:

> actress' costume, genius' test score, the three technicians' samples, Jesus' parables, the companies' joint project

■ *Single-Syllable Nouns Ending in "s"*

Form the possessive of single-syllable nouns ending in "s" or an "s" sound by adding an apostrophe plus "s." Examples:

> Hoss's horse, Rex's song, the boss's progress report

■ *All Nouns Not Ending in "s"*

Form the possessive of all singular and plural nouns not ending in "s" by adding an apostrophe plus "s." Examples:

> the man's hat, the men's team, the company's policy

■ *Paired Nouns*

Form the possessive of paired nouns by first determining whether there is joint ownership or individual ownership. For joint ownership, make only the last noun possessive. For individual ownership, make both nouns possessive. Examples:

> Susan and Terry's project was entered in the science fair, but Tom's and Scott's projects were not.

APT/LIABLE/LIKELY

Maintain the distinctions in these three similar words.

1. *Apt* is an adjective that means "appropriate," "suitable," or "has an aptitude for."
2. *Liable* is an adjective that means "legally obligated" or "subject to."
3. *Likely* is either an adjective that means "probable" or "promising" or an adverb that means "probably." As an adverb, it should be preceded by a qualifier such as *quite.*

Examples:

- The successful advertising campaign showed that she could select an *apt* phrase for selling products.

- Jonathan is *apt* at running good meetings. He always hands out an agenda and always ends on time.
- The contract makes clear who is *liable* for any on-site damage.
- Completing the warehouse without an inspection will make the contractor *liable* to lawsuits from the owner.
- A *likely* result of the investigation will be a change in the law. [*likely* as an adjective]
- The investigation will quite *likely* result in a change in the law. [*likely* as an adverb]

ASSURE/ENSURE/INSURE

As a verb meaning "to promise," *assure* is used in reference to people, as in "We want to assure you that our crews will strive to complete the project on time." *Assure* and its derivatives (like *assurance*) should be used with care in technical contexts, for these words can be viewed as a promise or guarantee.

The synonyms *ensure* and *insure* are verbs meaning to "make certain." Like *assure,* they imply a level of certainty that is not always appropriate in engineering and the sciences. When their use is deemed appropriate, the preferred word is *ensure;* reserve *insure* for sentences in which the context is insurance. Examples:

- I can *assure* you that our representatives will be on site to answer questions that the subcontractor may have.
- To *ensure* that the project stays within schedule, we are building in 10 extra days for bad weather. (An alternative that expresses less certainty and obligation: *So that* the project stays within schedule, we are building in 10 extra days for bad weather.)
- We *insured* the truck for $15,000.

AUGMENT/SUPPLEMENT

Augment is a verb that means "to increase in size, weight, number, or importance." *Supplement* is either (1) a verb that means to "add to" something to make it complete or to make up for a deficiency *or* (2) a noun that means "the thing that has been added." Examples:

- The power company supervisor decided to *augment* the line crews in five counties.

- He *supplemented* the deficient audit report by adding the three annual statements requested by the auditors.
- The three accounting *supplements* helped support the conclusions of the audit report.

AWHILE/A WHILE

Though similar in meaning, this pair is used differently. *Awhile* is an adverb that means "for a short time." Because *for* is already a part of its definition, the word cannot be preceded by the proposition *for*. The noun *while*, however, can be preceded by the two words *for a*, giving it essentially the same meaning as *awhile*. Examples:

- Kirk waited *awhile* before trying to restart the generator.
- Kirk waited *for a while* before trying to restart the generator.

BALANCE/REMAINDER/REST

Balance should be used as a synonym for *remainder* only in the context of financial affairs. *Remainder* and *rest* are synonyms to be used in other nonfinancial contexts. Examples:

- The account had a *balance* of $500, which was enough to avoid a service charge.
- The *remainder* [or *rest*, but not *balance*] of the day will be spent on training in oral presentations for proposals.
- During the *rest* [not *balance*] of the session, we learned about the new office equipment.

BECAUSE/SINCE

Maintain the distinction between these two words. *Because* establishes a cause-effect relationship, whereas *since* is associated with time. Examples:

- *Because* he left at 3 p.m., he was able to avoid rush hour.
- *Since* last week, her manufacturing team completed 3,000 units.

Note that *because* should not be used in the wordy construction "the reason is because." Examples:

- John wants the transfer *because* he needs the experience working at another office. [*not* "The reason John needs the transfer is because he needs the experience working at another office."]
- The ship sank *because* it was overloaded. [*not* "The reason the ship sank was because it was overloaded."]

BETWEEN/AMONG

The distinction between these two words has become somewhat blurred. However, many readers still prefer to see *between* used with reference to only two items and *among* used for three or more items. Examples:

- The agreement was just *between* my supervisor and me. No one else in the group knew about it.
- The proposal was circulated *among* all five members of the writing team.
- *Among* Sally, Todd, and Fran, there was little agreement about the long-term benefits of the project.

BI-/SEMI-/BIANNUAL/BIENNIAL

The prefixes *bi* and *semi* can cause confusion. Generally, *bi* means "every two years, months, weeks, etc.," whereas *semi* means "twice a year, month, week, etc." Yet many readers get confused by the difference, especially when they are confronted with a notable exception such as *biannual* (which means twice a year) and *biennial* (which means every two years).

Your goal, as always, is clarity for the reader. Therefore, it is best to write out meanings in clear prose, rather than relying on prefixes that may not be understood. Examples:

- We get paid twice a month [preferable to *semimonthly* or *biweekly*].
- The part-time editor submits articles every other month [preferable to *bimonthly*].
- We hold a company social gathering twice a year [preferable to *biannually* or *semi-annually*].
- The auditor inspects our safety files every two years [preferable to *biennially*].

BRACKETS

Use a pair of brackets (1) to set off explanatory material already contained within another parenthetical statement (see Parentheses entry in this appendix) or (2) to draw attention to a comment that you, as the writer, are making within a passage you are quoting. Examples:

> Two of our studies have shown that Colony Dam satisfies safety standards. (See Figure 4–3 [Dam Safety Record] for a complete record of our findings.) In addition, the county engineer, Greg Parker, has a letter on file that will give further assurance to homeowners on the lake. Parker notes the following in his letter: "After finishing my three-month study [Parker completed it in July 1995], I have concluded that the Colony Dam meets all required safety standards."

CAPITAL/CAPITOL

Capital is a noun whose main meanings are (1) a city or town that is a government center, (2) wealth or resources, or (3) net worth of a business or the investment that has been made in the business by owners. *Capital* can also be an adjective meaning (1) "excellent," (2) "primary," or (3) "related to the death penalty." Finally, *capital* can be a noun or adjective referring to upper-case letters.

Capitol is a noun or adjective that refers to a building where a legislature meets. With a capital letter, it refers exclusively to the building in Washington, D.C., where the U.S. Congress meets. Examples:

- The *capital* of Pickens County is Jasper, Georgia.
- Our family *capital* was reduced by the tornado and hurricane.
- She had invested significant *capital* in the carpet factory.
- Their proposal contained some *capital* ideas that would open new opportunities for our firm.
- In some countries, armed robbery is a *capital* offense.
- The students visited the *capitol* building in Atlanta. Next year they will visit the *Capitol* in Washington, D.C., where they will meet several members of Congress.

CAPITALIZATION

As a rule, you should capitalize names of specific people, places, and things—sometimes called "proper" nouns. For example, capitalize specific streets, towns, trademarks, geological eras, planets, groups of stars, days of the week, months of the year, names of organizations, holidays, and colleges. Nonspecific nouns—called "common" nouns—are not capitalized.

Of course, if it were this easy there wouldn't be so much confusion about capitalization. There is a tendency in business to overuse capitalization, as in the use of capitals with titles of positions. An example of improper usage follows: "The Director of Marketing visited our office last week." Excessive use of capitals appears pompous and is inappropriate in technical writing. The best practice today includes limited, judicious use of capitals. When in doubt, leave it lower case. Here are the most frequently capitalized items:

■ MAJOR WORDS IN TITLES OF BOOKS AND ARTICLES

Capitalize prepositions and articles only when they are the first words in titles. Examples:

- *For Whom the Bell Tolls*
- *In Search of Excellence*
- *The Power of Positive Thinking*

■ NAMES OF SPECIFIC PLACES AND GEOGRAPHICAL LOCATIONS

Examples:

- Washington Monument
- Rose Bowl
- Dallas, Texas
- Summit County, Ohio

■ NAMES OF AIRCRAFT AND SHIPS

Examples:

- Air Force One
- S.S. Arizona
- Nina, Pinta, and the Santa Maria

■ NAMES OF DEPARTMENTS, OFFICES, AND COMMITTEES IN AN ORGANIZATION

Examples:

- *Personnel Department*
- *International Division*
- *Benefits Committee*

However, when shorthand terms are used for these entities, capitals are not used. Example: Being an elected member of the Benefits Committee, I am concerned that the committee has not met in six months.

■ *TITLES THAT COME BEFORE NAMES*

Accepted practice calls for capitalizing titles only when they come directly before the name of the person who holds the title. When titles are used by themselves or follow a person's name, usually they are not capitalized. Examples:

- After breakfast, Chancellor Hairston opened the session.
- To our chagrin, Councilwoman Sharon Jones has served for nine years.
- We thought Professor Ginsberg gave a fine lecture.
- Jane Cannon, a professor in the Math Department, is on leave.
- Rob Smith, president of BGK, Inc., took a well-deserved vacation.
- Our company president was visiting the Singapore office.

The most noteworthy exceptions to this rule are the President of the United States and other heads of state; such titles are usually capitalized in any context. There are few other exceptions. If a company insists on breaking this rule—as in policy documents—it should capitalize ALL positions, not just those at the highest levels.

CENTER ON/REVOLVE AROUND

The key to using these phrases correctly is to think about their literal meaning. For example, you center *on* (not around) a goal, just as you would center on a target with a gun or bow and arrow. Likewise, your hobbies revolve *around* your early interest in water sports, just as the planets revolve around the sun in our solar system. Examples:

- All her selling points in the proposal *centered on* the need for greater productivity in the factory.
- At the latest annual meeting, some stockholders argued that most of the company's recent projects *revolved around* the CEO's interest in attracting attention from the media.

CITE/SITE/SIGHT

1. *Cite* is a verb meaning "to quote as an example, authority, or proof." It can also mean "to commend" or "to bring before a court of law" (as in receiving a traffic ticket).
2. *Site* usually is a noun that means "a particular location." It can also be a verb that means "to place at a location," as with a new school being sited by the town square, but this usage is not preferred. Instead use a more conventional verb, such as *built.*
3. *Sight* is a noun meaning "the act of seeing" or "something that is seen." Or it can be a verb meaning "to see or observe."

Examples:

- We *cited* a famous geologist in our report on the earthquake.
- Rene was *cited* during the ceremony for her exemplary service to the city of Roswell.
- The officer will *cite* the party-goers for disturbing the peace.
- Although five possible dorm *sites* were considered last year, the college administrators decided to build [preferred over *site*] the dorm at a different location.
- The *sight* of the flock of whooping cranes excited the visitors.
- Yesterday we *sighted* five whooping cranes at the marsh.

COLONS

Colons are quite common in technical writing because they are used with lists. Here are the three main ways they are used:

■ *AS A FORMAL BREAK BEFORE A LIST*

As mentioned in the Lists entry, you should place a colon immediately after the last word before a list. Examples:

- Our field study included the tasks listed below:
- In our field study we were asked to complete the following tasks:

Note that a complete clause precedes the colon. Avoid incomplete constructions such as "In our field study, we were asked to:"

■ *AS A FORMAL BREAK BEFORE A POINT OF CLARIFICATION OR ELABORATION*

Examples:

- They were interested in just one issue: quality.
- John loved the dish because of its main ingredient: salmon.

■ *AS A FORMAL BREAK BEFORE A SERIES WITHIN A SENTENCE*

There are two ways to lead into a series within a sentence. First, you can lead smoothly into the series without a break and thus without using a colon. Second, you can include a formal stop that does require a colon. Both options are shown below.

- Next month I will attend meetings in four cities: Houston, Austin, Laredo, and Abilene.
- Next month I will attend conferences in Houston, Austin, Laredo, and Abilene.

COMMAS

Many writers struggle with commas, and for good reason. First, the teaching of punctuation is approached in different, and sometimes contradictory, ways. Second, comma rules are subject to some interpretation. And third, problems with commas sometimes reflect more fundamental problems with sentence structure.

The best strategy is to memorize a few basic rules of comma use. This knowledge will give you confidence in your ability to handle the mechanics of editing. (If you need help understanding any grammatical terms used in the rules, like *compound sentence,* refer to the Sentence Structure entries in this appendix.)

■ COMMAS IN A SERIES

Use commas to separate words, phrases, and short clauses written in a series of three or more items. According to current U.S. usage, a comma always precedes the *and* that comes before the last item in a series. Examples:

- The samples contained gray sand, sandy clay, and silty sand.
- The meal included steak, potatoes, and peaches and cream.

In the second example, note that the *and* before the last item in the series (and thus the one that is punctuated) is the one that precedes *peaches.* The other *and*—the one that precedes *cream*—has no comma before it because it only connects two parts of the last element in the series.

■ COMMAS IN COMPOUND SENTENCES

Use a comma before the conjunction that joins main clauses in a compound sentence. Example:

> We finished drilling the well, and then we grouted the holes with concrete.

The comma is used here because it separates two complete clauses, each with its own subject and verb ("we finished" and "we grouted"). If the second *we* had been deleted, there would be only one clause containing one subject and two verbs ("we finished and grouted"). Thus no comma would be needed. Of course, be certain that you don't use this comma rule to string together intolerably long sentences. Divide long sentences when you can. (See the Sentence Structure entry.)

■ COMMAS WITH NONESSENTIAL MODIFIERS

Set off nonessential modifiers with commas—either at the beginning, middle, or end of sentences. Nonessential modifiers are phrases that add more information to a sentence, rather than greatly changing its meaning. When you speak, often there is a pause between this kind of modifier and the main part of the sentence, giving a clue that a comma break is needed. Examples:

- The report, which we submitted three weeks ago, indicated that the company would not be responsible for transporting hazardous wastes.
- The report that we submitted three weeks ago indicated that the company would not be responsible for transporting hazardous wastes.

The first example includes a nonessential modifier. It would be spoken with pauses and therefore uses separating commas. The second example includes an essential modifier. It would be spoken without pauses and therefore includes no separating commas.

■ COMMAS WITH ADJECTIVES IN A SERIES

Use a comma to separate two or more adjectives that modify the same noun. To help you decide if adjectives modify the same noun, use this test: if you can reverse their position and still retain the same meaning, then the adjectives modify the same word and should be separated by a comma. Example:

> Jason opened the two containers in a dry, well-lighted place.

■ COMMAS WITH INTRODUCTORY ELEMENTS

Use a comma after introductory phrases or clauses of about five words or more. Example:

> After completing its topographic survey of the area, the crew returned to headquarters for the weekly project meeting.

Commas like the one after *area* help readers separate secondary or modifying points from the main idea, which of course should be in the main clause. Without these commas, there may be difficulty reading the sentence properly.

■ COMMAS IN LISTS

Use commas in lists that you wish to treat as complete sentences. (See the Lists entries for a complete discussion of use and punctuation of listings.)

◼ *COMMAS IN DATES, TITLES, ETC.*

Abide by the conventions of comma usage in punctuating dates, titles, geographic place names, and addresses. Examples:

- May 3, 1999, is the projected date of completion.
- John F. Dunwoody, P.E., has been hired to direct the project because he is a registered engineer.
- McDuff, Inc., has been selected.
- He listed Dayton, Ohio, as his permanent residence.

Note the need for commas after the year "1999," the designation "Inc.," the state name "Ohio," and the title "P.E." (as there would be for Ph.D.).

COMPLEMENT/COMPLIMENT

Both words can be nouns and verbs, and both have adjective forms (*complementary, complimentary*).

COMPLEMENT

This word is used as a noun to mean "that which has made something whole or complete," as a verb to mean "to make whole, to make complete," or as an adjective to mean "serving as a complement." You may find it easier to remember the word by recalling its mathematical definition: two complementary angles must always equal 90 degrees. Examples:

- (as noun): The *complement* of five technicians brought our crew strength up to 100 percent.
- (as verb): The firm in Canada *complemented* ours in that together we won a joint contract for work in both countries.
- (as adjective): Seeing the project manager and her secretary work so well together made clear their *complementary* relationship in getting office work done.

COMPLIMENT

This word is used as a noun to mean "an act of praise, flattery, or admiration," as a verb to mean "to praise, to flatter," or as an adjective to mean "related to praise or flattery, or without charge." Examples:

- (as noun): He appreciated the verbal *compliments,* but he also hoped they would result in a substantial raise.
- (as verb): Howard *complimented* the crew for finishing the job on time and within budget.

- (as adjective): We were fortunate to receive several *complimentary* copies of the new software from the publisher.

COMPOSE/COMPRISE

Compose means "to make up or be included in," whereas *comprise* means "to include or consist of." The easiest way to remember this relationship is to memorize one sentence:

> The parts *compose* the whole, but the whole *comprises* the parts.

One more point to remember: the common phrase "is comprised of" is a substandard, unacceptable replacement for "comprise" or "is composed of." Careful writers don't use it. Examples:

- Seven separate layers *compose* the soils that were uncovered at the site.
- The borings revealed a stratigraphy that *comprises* [not *is comprised of*] seven separate layers.

CONSUL/COUNCIL/COUNSEL

These three words can be distinguished by meaning and, in part, by their use within a sentence.

> *Consul:* A noun meaning an official of a country who is sent to represent that country's interests in a foreign land.
>
> *Council:* A noun meaning an official group or committee.
>
> *Counsel:* A noun meaning an adviser or advice given OR a verb meaning to provide advice.

Examples:

- (consul) The Brazilian consul met with consular officials from three other countries.
- (council) The Human Resources Council of our company recommended a new retirement plan to the company president.
- (counsel—as noun) After the tragedy, they received legal counsel from their family attorney and spiritual counsel from their minister.
- (counsel—as verb) As a communications specialist, Roberta helps to counsel employees who are involved in various types of disputes.

CONTINUOUS/CONTINUAL

These words have close yet distinctly different meanings. *Continuous* and *continuously* should be used in reference to uninterrupted, unceasing activities. *Continual* and *continually* should be used with activities that are intermittent, or repeated at intervals. If you think your reader may not understand the difference, use synonyms that will be more clear—such as *uninterrupted* for *continuous* and *intermittent* for *continual*. Examples:

- Because it rained *continuously* from 10 a.m. until noon, we were unable to find even a minute of clear sky to move our equipment onto the property.
- We *continually* checked the water pressure for three days before the equipment arrived, while also using the time to set up tests.

CRITERION/CRITERIA

Coming from the Latin language, *criterion* and *criteria* are the singular and plural forms of a word that means "rationale or reasons for selecting a person, place, thing, or idea." A common error is to use *criteria* as both a singular and plural form, but such misuse disregards a distinction recognized by many readers. Maintain the distinction in your writing. Examples:

- Among all the qualifications we established for the new position, the most important *criterion* for success is good communication skills.
- She had to satisfy many *criteria* before being accepted into the honorary society of her profession.

DATA/DATUM

Coming from the Latin, the word *data* is the plural form of *datum*. Traditionalists in the engineering and scientific community still consider *data* a plural form in all cases. However, in recent years some writers have considered *data* a plural form in some cases, when it means "facts," and a singular form in other cases, when it means "information." This book suggests you maintain the traditional use of data as a plural form, unless you have good reason to do otherwise. Examples:

- These *data* show that there is a strong case for building the dam at the other location.
- This particular *datum* shows we need to reconsider recommendations put forth in the original report.

If you consider the traditional singular form—*datum*—to be awkward, use substitutes such as "This item in the data demonstrates our point" or "One of the data demonstrates our point." Singular subjects like *one* or *item* allow you to keep your original meaning without using the word *datum*.

DEFINITE/DEFINITIVE

Though similar in meaning, these two words have slightly different contexts. *Definite* refers to that which is precise, explicit, or final. *Definitive* has the more restrictive meaning of "authoritative" or "final." Examples:

- It is now *definite* that he will be assigned to the London office for six months.
- He received the *definitive* study on the effect of the oil spill on the marine ecology.

DISCRETE/DISCREET/DISCRETION

The adjective *discrete* suggests something that is separate, or something that is made up of many separate parts. The adjective *discreet* is associated with actions that require caution, modesty, or reserve. The noun *discretion* refers to the quality of being "discreet" or the freedom a person has to act on his or her own. Examples:

- The company orientation program includes a writing seminar, which is a *discrete* training unit offered for one full day.
- The orientation program includes five *discrete* units.
- As a member of the human resources staff, Sharon was *discreet* in her handling of personal information about employees.
- Every human resources employee was instructed to show *discretion* in handling personal information about employees.
- By starting a "flextime" program, the company will give employees a good deal of *discretion* in selecting the time to start and end their workday.

DISINTERESTED/UNINTERESTED

These words have quite different meanings. *Disinterested* means "without prejudice or bias," whereas *uninterested* means "showing no interest." Examples:

- The agency sought a *disinterested* observer who had no stake in the outcome of the trial.
- They spent several days talking to officials from Iceland, but they still remain *uninterested* in performing work in that country.

DUE TO/BECAUSE OF

Mixing these two phrases can also cause confusion. *Due to* is an adjective phrase meaning "attributable to." It usually follows a "to be" verb such as *is, was,* or *were.* It should not be used in place of prepositional phrases such as *because of, owing to,* or *as a result of.* Examples:

■ The cracked walls were *due to* the lack of proper fill being used during construction.
■ We won the contract *because of* [not *due to*] our thorough understanding of the request for proposal.

EACH OTHER/ONE ANOTHER

Each other occurs in contexts that include only two persons, whereas *one another* occurs in contexts that include three or more persons. Examples:

■ Shana and Katie worked closely with *each other* during the project.
■ All six members of the team conversed with *one another* regularly through email.

E.G./I.E.

The abbreviation *e.g.* means "for example," whereas *i.e.* means "that is." These two Latin abbreviations are often confused. Thus many writers prefer to write them out, rather than risk confusion on the part of the reader. Examples:

■ He visited many cities where his firm plans to open offices—e.g., [or, preferably, *for example*], Kansas City, New Orleans, and Seattle.
■ A spot along the Zayante Fault was the earthquake's epicenter—i.e., [or, preferably, *that is*], the focal point for seismic activity.

ENGLISH AS A SECOND LANGUAGE (ESL)

Technical writing challenges native English speakers and non-native speakers alike. The purpose of this section is to present a basic description of three grammatical forms: articles, prepositions, and verbs. These forms may require more intense consideration from international students when they complete technical writing assignments. Each issue is described using the ease-of-operation section from the memo in Figure 2–2 on page 22. The passage, descriptions, and charts work together to show how these grammar issues function collectively to create meaning.

Ease of Operation—Article Usage

The AIM 500 is so easy to operate that **a** novice can learn to transmit **a** document to another location in about two minutes. Here's **the** basic procedure:

1. Press **the** button marked TEL on **the** face of **the** fax machine. You then hear **a** dial tone.
2. Press **the** telephone number of **the** person receiving **the** fax on **the** number pad on **the** face of **the** machine.
3. Lay **the** document facedown on **the** tray at **the** back of **the** machine.

At this point, just wait for **the** document to be transmitted—about 18 seconds per page to transmit. **The** fax machine will even signal **the** user with **a** beep and **a** message on its LCD display when **the** document has been transmitted. Other more advanced operations are equally simple to use and require little training. Provided with **the** machine are two different charts that illustrate **the** machine's main functions.

The size of **the** AIM 500 makes it easy to set up almost anywhere in **an** office. **The** dimensions are 13 inches in width, 15 inches in length, and 9.5 inches in height. **The** narrow width, in particular, allows **the** machine to fit on most desks, file cabinets, or shelves.

▪ ARTICLES

Articles are one of the most difficult forms of English grammar for non-native English speakers, mainly because some language systems do not use them. Thus speakers of particular languages may have to work hard to incorporate the English article system into their language proficiency.

The English articles include *a, an,* and *the.*

- *A* and *an* express indefinite meaning when they refer to nouns or pronouns or pronouns that are not specific. The writer believes the reader does not know the noun or pronoun.
- *The* expresses definite meaning when it refers to a specific noun or pronoun. The writer believes the reader knows the specific noun or pronoun.

ESL writers choose the correct article only when they (1) know the context or meaning, (2) determine whether they share information about the noun with the reader, and (3) consider the type of noun following the article.

The ease-of-operation passage includes 32 articles that represent two types—definite and indefinite. When a writer and a reader share knowledge of a noun, the definite article should be used. On 26 occasions the articles in the passage suggest the writer and reader share some knowledge of a count noun. Count nouns are nouns that can be counted (pen, cloud, memo). Examples of non-count nouns are sugar, air, and beef.

For example, the memo writer and the memo recipient share knowledge of the particular model fax machine—the AIM 500. Thus, "the" is definite when it refers to "the fax machine" in the memo. Notice, however, that "document" becomes definite only after the second time it is mentioned ("Lay the document face down. . . ."). In the first reference to *document, a* document refers to a document about which the writer and reader share no knowledge. The memo writer cannot know which document the reader will fax. Only in the second reference do the writer and reader know the document to be the one the reader will fax.

The indefinite article *a* occurs five times, while *an* occurs once. Each occurrence signals a singular count noun. The reader and the writer share no knowledge of the nouns that follow the *a* or *an,* so an indefinite article is

Articles from "Ease of Operation" Excerpt

Article	Noun	Type	Comment
The	AIM 500	definite	first mention—shared knowledge
a	novice	indefinite	first mention—no shared knowledge
a	document	indefinite	first mention—no shared knowledge
the	basic procedure	definite	
the	button	definite	
the	face	definite	
the	fax machine	definite	first mention without proper name, with reader/writer shared knowledge
a	dial tone	indefinite	first mention—no shared knowledge
the	telephone number	definite	
the	person	definite	
the	fax	definite	
the	number pad	definite	
the	face	definite	
the	machine	definite	
the	document	definite	
the	tray	definite	
the	back	definite	
the	machine	definite	
the	document	definite	second mention
the	fax machine	definite	
the	user	definite	
a	beep	indefinite	first mention—no shared knowledge
a	message	indefinite	first mention—no shared knowledge
the	document	definite	
the	machine	definite	
the	machine's main functions	definite	
the	size	definite	
The	AIM 500	definite	
an	office	indefinite	first mention—preceding vowel sound—no shared knowledge
the	dimensions	definite	
the	narrow width	definite	
the	machine	definite	

appropriate. *A* precedes nouns beginning with consonant sounds. *An* precedes nouns beginning with vowel sounds. Indefinite articles seldom precede non-count nouns unless a non-count functions as a modifier (a beef shortage).

Definite and indefinite articles are used more frequently than other articles; however, other articles do exist. The "generic" article refers to classes or groups of people, objects, and ideas. If the fax machine is thought of in a general sense, the meaning changes. For example, "the fax machine increased office productivity by 33%." *The* now has a generic meaning representing fax machines in general. The same generic meaning can apply to the plural noun, but such generic use requires no article. "Fax machines increased office productivity by 33%." *The* in this instance is a generic article.

Ease of Operation—Verb Usage

The AIM 500 is so easy to operate that a novice **can learn** to transmit a document to another location in about two minutes. Here's the basic procedure:

1. **Press** the button marked TEL on the face of the fax machine. You then **hear** a dial tone.
2. **Press** the telephone number of the person receiving the fax on the number pad on the face of the machine.
3. **Lay** the document facedown on the tray at the back of the machine.

At this point, just **wait** for the document to be transmitted—about 18 seconds per page to transmit. The fax machine **will** even **signal** the user with a beep and a message on its LCD display when the document **has been transmitted.** Other more advanced operations **are** equally simple to use and **require** little training. **Provided** with the machine **are** two different charts that **illustrate** the machine's main functions.

The size of the AIM 500 **makes** it easy to set up almost anywhere in an office. The dimensions **are** 13 inches in width, 15 inches in length, and 9.5 inches in height. The narrow width, in particular, **allows** the machine to fit on most desks, file cabinets, or shelves.

■ VERBS

Verbs express time in three ways—simple present, simple past, and future. *Wait, waited,* and *will wait* and *lay* (*to put*), *laid,* and *will lay* are examples of simple present, simple past, and future tense verbs. Unfortunately, the English verb system is more complicated than that. Verbs express more than time.

Verbs in the English language system appear either as regular or irregular forms. Regular verbs follow a predictable pattern.

Regular verbs

present	past	future
learn	learned	will learn
wait	waited	will wait
press	pressed	will press
signal	signaled	will signal
require	required	will require
provide	provided	will provide
illustrate	illustrated	will illustrate
allow	allowed	will allow
present perfect	**past perfect**	**future perfect**
have learned	had learned	will have learned
have waited	had waited	will have waited
have pressed	had pressed	will have pressed
have signaled	had signaled	will have signaled
have required	had required	will have required
have provided	had provided	will have provided
have illustrated	had illustrated	will have illustrated
have allowed	had allowed	will have allowed

- The form of the simple present tense verbs (*walk*) changes to the simple past tense with the addition of *ed* (*walked*).
- The past tense *walked* changes to the perfect tense with the addition of an auxiliary (helping) word to the simple past tense form. For example, "I have walked" is present perfect, and "I had walked" is past perfect.

  Irregular verbs do not follow a predictable pattern.

- Most importantly, the past tense is not created by adding *ed*. The simple present tense of lay (*to put*) changes completely in the simple past (*laid*).
- Like regular verbs, the past tense (laid) changes to the perfect tense with the addition of an auxiliary verb.

"I have laid the document facedown on the tray" is present perfect, and "I had laid the document facedown on the tray" is past perfect.

Let's examine four specific verb forms in the "Ease of Operation" passage.

1. *Is* represents a being or linking verb in the passage. Being verbs suggest an aspect of an experience or being (existence); for example, "He is still here," and "the fax is broken." Linking verbs connect a subject to a complement (completer); for example, "The fax machine is inexpensive."

Irregular verbs

present	past	future
is	was	will be
are	were	will be
hear	heard	will hear
do	did	will do
get	got	will get
see	saw	will see
write	wrote	will write
speak	spoke	will speak
present perfect	**past perfect**	**future perfect**
have been (he, she, it)	had been	will have been
have been (they)	had been	will have been
have heard	had heard	will have heard
have done	had done	will have done
have gotten	had gotten	will have gotten
have seen	had seen	will have seen
have written	had written	will have written
have spoken	had spoken	will have spoken

2. *Can learn* is the present tense verb *learn* preceded by a modal. Modals assist verbs to convey meaning. *Can* suggests ability or possibility. Other modals and their meanings appear below.

will	would	could	shall	should	might	must
scientific fact possibility determination	hypothetical	hypothetical	formal will	expectation obligation	possibility	necessity

3. *Here's* shows a linking verb (*is*) connected to its complement (here). The sentence in its usual order—subject first followed by the verb—would appear as, "The basic procedure is here." Article—adjective—noun—linking verb—complement.
4. *Press, Lay,* and *wait* (for) share at least four common traits: present tense, singular number, action to transitive, and understood subject of *you.*

Though *you* does not appear in the text, the procedure clearly instructs the person operating the fax machine—*you*. Action or transitive verbs express movement, activity, and momentum, and may take objects. Objects answer the questions who? what? to whom? or for whom? in relation to transitive verbs. For example, "Press the button." "Hear a dial tone." "Press the telephone number." "Lay the document face down." Press what? Hear what? Lay what?

Verbs from "Ease of Operation" Excerpt

Verb	Tense	Number	Other details
is	present	singular	linking/being (is, was, been)
can learn	present	singular	*can* is a modal auxiliary implying "possibility"
here's (is)	present	singular	linking/being
Press	present	singular	understood *you* as subject
hear	present	singular	action/transitive
Press	present	singular	understood *you* as subject
Lay	present	singular	irregular (lay, laid, laid) singular—understood *you* as subject
wait	present	singular	understood *you* as subject
will signal	future	singular	action to happen or condition to experience—understood *you* as subject
has been transmitted	present perfect	singular	passive voice—action that began in the past and continues to the present
are	present	plural	linking/being
require	present	plural	action/transitive
Provided are	present perfect	plural	passive voice—action that began in the past and continues to the present
illustrate	present	plural	action/transitive
makes	present	singular	action/transitive
are	present	plural	linking/being
allows	present	singular	action/transitive

Ease of Operation—Preposition Usage

The AIM 500 is so easy to operate that a novice can learn to transmit a document **to** another location **in** about two minutes. Here's the basic procedure:

1. Press the button marked TEL **on** the face **of** the fax machine. You then hear a dial tone.

2. Press the telephone number **of** the person receiving the fax **on** the number pad **on** the face **of** the machine.
3. Lay the document facedown **on** the tray **at** the back **of** the machine.

At this point, just wait **for** the document to be transmitted—**about** 18 seconds **per** page to transmit. The fax machine will even signal the user **with** a beep and a message **on** its LCD display when the document has been transmitted. Other more advanced operations are equally simple to use and require little training. Provided **with** the machine are two different charts that illustrate the machine's main functions.

The size **of** the AIM 500 makes it easy to set up almost anywhere **in** an office. The dimensions are 13 inches **in** width, 15 inches **in** length, and 9.5 inches **in** height. The narrow width, **in** particular, allows the machine to fit **on** most desks, file cabinets, or shelves.

■ *PREPOSITIONS*

Prepositions are words that become a part of a phrase composed of the preposition, a noun or pronoun, and any modifiers. Notice the relationships expressed within the prepositional phrases and the ways they affect meaning in the sentences. In the "Ease of Operation" passage, about half the prepositional phrases function as adverbs noting place or time. The other half function as adjectives.

Place	or	Location	Time
at		on	before
in		above	after
below		around	since
beneath		out	during
over		underneath	
within		under	
outside		near	
into		inside	

One important exception is a preposition that connects to a verb to make a "prepositional" verb, *wait for*. Another interesting quality of prepositions is that sometimes more than one can be used to express similar meaning. In the "Ease of Operation" passage, for example, both *on* the tray and *at* the back indicate position. Another way to state the same information is *on* the tray *on* the back.

Prepositions from "Ease of Operation" Excerpt

Preposition	Noun Phrase	Comment
to	another location	direction toward
in	(about) two minutes	approximation of time
on	the face	position
of	the fax machine	originating at or from
of	the person	associated with
on	the number	position
on	the face	position
of	the machine	originating at
on	the tray	position
at	the back	position of
of	the machine	originating at
At	this point	on or near the time
for	the document	indication of object of desire
per	page	for every
about	18 seconds	adverb = approximation
with	a beep and a message	accompanying
on	its LCD display	position
with	the machine	accompanying
of	the AIM 500	originating at or from
in	an office	within the area
in	width	with reference to
in	length	with reference to
in	height	with reference to
in	particular	with reference to
on	most desks, file cabinets, or shelves	position

FARTHER/FURTHER

Though similar in meaning, these two words are used differently. *Farther* refers to actual physical distance, whereas *further* refers to non-physical distance or can mean "additional." Examples:

- The overhead projector was moved *farther* from the screen so that the print would be easier to see.
- *Farther* up the old lumber road, they found footprints of an unidentified mammal.

- As he read *further* along in the report, he began to understand the complexity of the project.
- She gave *further* instructions after they arrived at the site.

FEWER/LESS

The adjective *fewer* is used before items that can be counted, whereas the adjective *less* is used before mass quantities. When errors occur, they usually result from *less* being used with countable items, as in this *incorrect* sentence: "We can complete the job with less men at the site." Examples:

- "The newly certified industrial hygienist signed with us because the other firm in which he was interested offered *fewer* [not *less*] benefits."
- "There was *less* sand in the sample taken from 15 ft than in the one taken from 10 ft."

FLAMMABLE/INFLAMMABLE/NONFLAMMABLE

Given the importance of these words in avoiding injury and death, make sure to use them correctly—especially in instructions.

1. *Flammable* means "capable of burning quickly" and is accepted usage.
2. *Inflammable* has the same meaning as *flammable*. However, it is *not* acceptable usage because readers may confuse it with *nonflammable*.
3. *Nonflammable* means "not capable of burning" and is accepted usage.

Examples:

- They marked the package *flammable* because its contents could be easily ignited by a spark. [Note that *flammable* is preferred here over its synonym, *inflammable*.]
- The supervisor was not worried about placing the crates near the heating unit, because all the contents were *nonflammable*.

FORMER/LATTER

These two words direct the reader's attention to previous points or items. *Former* refers to that which came first, whereas *latter* refers to that which came last. Note that the words are used together only with *two* items or points—*not* with three or more. Also, you should know that some readers may prefer you avoid *former* and *latter* altogether, for the construction may force them to look back to previous sentences to understand your meaning. The second example below gives an alternative.

- (with former/latter) The airline's machinists and flight attendants went on strike yesterday. The *former* left work in the morning, whereas the *latter* left work in the afternoon.
- (without former/latter) The airline's machinists and flight attendants went on strike yesterday. The machinists left work in the morning, whereas the flight attendants left work in the afternoon.

FORTUITOUS/FORTUNATE

The adjective *fortuitous* indicates an action is unexpected, whether it is desirable or not. The adjective *fortunate* indicates an action is desired. A common error is the wrong assumption that "fortuitous" events must also be "fortunate." Examples:

- Seeing McDuff's London manager at the conference was quite *fortuitous*, for I had not been told that he also was attending.
- It was indeed *fortunate* that I encountered the Tokyo manager, for we had the chance to talk about a project involving both our offices.

GENERALLY/TYPICALLY/USUALLY

Words like these can be useful qualifiers in your reports. All three indicate that something is often, but not always, true. Place these adverb modifiers as close as possible to the words they modify. For example, in the first example below it would be inaccurate to write "were typically sampled," because "typically" modifies the entire verb phrase "were sampled." Examples:

- Cohesionless soils *typically* were sampled by driving a 2-in. diameter, split-barrel sampler. (Active-voice alternative: *Typically,* we sampled cohesionless soils by driving a 2-in. diameter, split-barrel sampler.)
- For projects like the one you propose, *usually* the technician will clean the equipment before returning to the office.
- It *generally* is known that sites for dumping waste should be equipped with appropriate liners.

GOOD/WELL

Though these words have similar meanings, *good* is used as an adjective, and *well* is used as an adverb. A common usage error occurs when writers use the adjective when the adverb is required. Examples:

- It is *good* practice to submit three-year plans on time.
- Although he had a slight case of the flu, he felt *well* enough to attend the seminar.

HOPEFULLY

Use and abuse of this adverb have attracted far more attention than it probably deserves. Yet you should know that some readers will make judgments about your level of literacy by your ability to use *hopefully* correctly. Use the word correctly to avoid drawing attention away from the content of your writing.

Hopefully is an adverb that means "with hope," *not* a one-word substitute for "I hope that." In other words, use *hopefully* only when it serves as an adverb and explicitly means "with hope." Examples:

- I looked *hopefully* through the files for the client's lost report. [That is, I looked "with hope."]
- They waited *hopefully* for the results of the proposal competition. [That is, they waited "with hope."]
- *I hope* I'll find the report before my client arrives for the meeting. [Not "Hopefully, I'll find . . ."]
- *They hoped* their firm would win the proposal competition. [Not "Hopefully, their firm would win . . ."]

IMPLY/INFER

The person doing the speaking or writing "implies." The person hearing or reading the words "infers." Thus the word *imply* requires an active role, whereas the word *infer* requires a passive role. When you imply a point, your words suggest rather than state a point. When you infer a point, you form a conclusion or deduce meaning from someone else's words. Examples:

- The contracts officer *implied* that there would be stiff competition for that $20 million waste treatment project.
- We *inferred* from her remarks that the firm selected for the work must have completed similar projects recently.

ITS/IT'S

These words are often confused. *It's* with the apostrophe is *only* used as a contraction for "it is" or "it has." The other form—*its*—is a possessive pronoun. (*Its'* is not a word.) Examples:

- Because of the rain, *it's* [or *it is*] going to be difficult to move the equipment to the site.
- *It's* [or *it has*] been a long time since we submitted the proposal.
- The company completed *its* part of the agreement on time.

LAY/LIE

These two verbs are troublesome, and you need to know some basic grammar to use them correctly.

1. *Lay* means "to place." It is a transitive verb; thus it takes a direct object to which it conveys action. ("She laid down the printout before starting the meeting.") Its main forms are *lay* (present), *laid* (past), *laid* (past participle), and *laying* (present participle).
2. *Lie* means "to be in a reclining position." It is an intransitive verb; thus it does not take a direct object. ("In some countries, it is acceptable for workers to lie down for a mid-day nap.") Its main forms are *lie* (present), *lay* (past), *lain* (past participle), and *lying* (present participle).

If you want to use these words with confidence, remember the transitive/intransitive distinction and memorize the principal parts. Examples:

- (lay) I will *lay* the notebook on the lab desk before noon.
- (lay) I have *laid* the notebook there before.
- (lay) I was *laying* the notebook down when the phone rang.
- (lie) The watchdog *lies* motionless at the warehouse gate.
- (lie) The dog *lay* there yesterday too.
- (lie) The dog has *lain* there for three hours today and no doubt will be *lying* there when I return from lunch.

LEAD/LED

Lead is either a noun that names the metallic element or a verb that means "to direct or show the way." *Led* is only a verb form, the past tense of the verb *lead*. Examples:

- The company bought rights to mine *lead* on the land.
- They chose a new president to *lead* the firm into the twenty-first century.
- They were *led* to believe that salary raises would be high this year.

LETTER FORMATS

Many organizations adopt letter formats used uniformly by all employees. Other organizations permit flexibility. In the latter case, you have three basic formats from which to choose—(1) block (with or without indented paragraphs), (2) modified block, or (3) simplified. Figures B–1, B–2, and B–3 provide guidelines for each format.

LIKE/AS

These two words are different parts of speech and thus are used differently in sentences. *Like* is a preposition and therefore is followed by an object—not an entire clause. *As* is a conjunction and thus is followed by a group of words that include a verb. *As if* and *as though* are related conjunctions. Examples:

- Gary looks *like* his father.
- Managers *like* John will be promoted quickly.
- If Teresa writes this report *as* she wrote the last one, our clients will be pleased.
- Our proposals are brief, *as* they should be.
- Our branch manager talks *as though* [or *as if*] the merger will take place soon.

LISTS: GENERAL POINTERS

Lists draw attention to three or more pieces of like information that are important enough to remove from standard text format. In other words, they are an attention-getting strategy. Following are some general pointers for using lists:

■ STRIVE FOR A RANGE OF FROM THREE TO NINE ITEMS

Most readers hold up to nine items in short-term memory. More than nine items gives the appearance of a disorganized "laundry list"; it may confuse

Letterhead of your organization

Two or more blank lines (adjust space to center letter on page)

Date of letter

Two or more blank lines (adjust space to center letter on page)

Address of reader

One blank line

Greeting

One blank line

Paragraph: single-spaced (indenting optional)

One blank line

Paragraph: single-spaced (indenting optional)

One blank line

Paragraph: single-spaced (indenting optional)

One blank line

Complimentary close

Three blank lines (for signature)

Typed name and title

One blank line

Typist's initials (optional: Writer's initials before typist's initials)

Computer file # (if applicable)

One blank line (optional)

Enclosure notation

One blank line (optional)

Copy notation

FIGURE B–1
Letter format: Block

Source: Technical Writing: A Practical Approach, 4th ed, (p. 238) by W. S. Pfeiffer, 2000, Upper Saddle River, NJ: Prentice Hall. Reprinted by permission.

| Letterhead of your organization |

Two or more blank lines (adjust space to center letter on page)

| Date of letter |

Two or more blank lines (adjust space to center letter on page)

| Address of reader |

One blank line

| Greeting |

One blank line

| Paragraph: single-spaced, with first line indented 5 spaces |

One blank line

| Paragraph: single-spaced, with first line indented 5 spaces |

One blank line

| Paragraph: single-spaced, with first line indented 5 spaces |

One blank line

| Complimentary close |

Three blank lines (for signature)

| Typed name and title |

One blank line

| Typist's initials (optional: Writer's initials before typist's initials) |

| Computer file # (if applicable) |

One blank line (optional)

| Enclosure notation |

One blank line (optional)

| Copy notation |

FIGURE B–2
Letter format: Modified block
Source: Technical Writing: A Practical Approach, 4th ed, (p. 239) by W. S. Pfeiffer, 2000, Upper Saddle River, NJ: Prentice Hall. Reprinted by permission.

Letterhead of organization

Two or more blank lines (adjust space to center letter on page)

Date of letter

Two or more blank lines (adjust space to center letter on page)

Address of reader

Three blank lines

Short subject line

Three blank lines

Paragraph: single-spaced, no indenting

One blank line

Paragraph: single-spaced, no indenting

One blank line

Paragraph: single-spaced, no indenting

Five blank lines (for signature)

Typed name and title

One blank line (optional)

Typist's initials (optional: Writer's initials before typist's initials)

Computer file # (if applicable)

One blank line (optional)

Enclosure notation

One blank line (optional)

Copy notation

FIGURE B–3
Letter format: Simplified

Source: Technical Writing: A Practical Approach, 4th ed, (p. 240) by W. S. Pfeiffer, 2000, Upper Saddle River, NJ: Prentice Hall. Reprinted by permission.

rather than clarify an issue. If you have more than nine items, consider grouping them in three or four categories, as you would in an outline. Groups make it easier for readers to grasp and remember many points.

■ USE BULLETS (•) FOR GROUPINGS THAT DO NOT INVOLVE SEQUENCE

The single round bullets are best. Avoid strange-shaped bullets that draw more attention to themselves than to your ideas.

■ USE NUMBERS TO CONVEY SEQUENCE

Numbers work best when you are listing ordered items such as steps, procedures, or ranked alternatives.

■ KEEP POINTS PARALLEL

Parallel means that all points in the same list are in the same grammatical form—whether a complete sentence, verb phrase, or noun phrase. If you change form in the midst of a listing, you risk upsetting the flow of your text. Example:

> To complete this project, we planned the following tasks:
>
> ■ Survey the site
> ■ Take samples from the three boring locations
> ■ Test selected samples in our lab
> ■ Report on the results of the study

The items above are in verb form (note introductory words *survey, take, test, report*). An alternative would be to put them in noun form, with a different lead-in.

LISTS: PUNCTUATION

You have three main options for punctuating lists. In all three you should (1) precede the list with a lead-in that is a complete thought, as in the preceding example, (2) place a colon after the last word of the lead-in, and (3) capitalize the first letter of the first word of each listed item.

■ *OPTION A: PLACE NO PUNCTUATION AFTER LISTED ITEMS*

This style is particularly appropriate when a list includes only short phrases. More and more writers are choosing this option. Example:

In this study we will develop recommendations that address the following concerns in your project:

- Site preparation
- Foundation design
- Sanitary sewer design
- Storm sewer design
- Geologic surface faulting
- Projections for regional land subsidence

■ *OPTION B: TREAT THE LIST LIKE A SENTENCE SERIES*

In this case you place commas or semicolons between items and a period at the end of the series. Whether you choose Option A or B largely depends on your own style. Example:

In this study we developed recommendations that dealt with four topics:

- Site preparation,
- Foundation design,
- Sewer construction, and
- Geologic faulting.

This option requires that you place an *and* after the comma that appears before the last item. Another variation of Option B occurs when you have internal commas within one or more of the items. In this case, you need to change the commas that follow the listed items into semicolons. Yet you still keep the *and* before the last item. Example:

Last month we completed environmental assessments at three sites:

- A gas refinery in Dallas, Texas;
- A chemical plant in Little Rock, Arkansas; and
- A waste pit outside of Baton Rouge, Louisiana.

■ OPTION C: TREAT EACH ITEM LIKE A SEPARATE SENTENCE

When items in a list are complete sentences, you can punctuate each one as a separate sentence, placing a period at the end of each. Example:

The main conclusions of our preliminary assessment are summarized here:

■ At five of the six borehole locations, petroleum hydrocarbons were detected at concentrations greater than a background concentration of 10 mg/kg.

■ No PCB concentrations were detected in the subsurface soils we analyzed. We will continue the testing, as outlined in our proposal.

■ Sampling and testing should be restarted three weeks from the date of this report.

LOOSE/LOSE

Loose, which rhymes with *goose,* is an adjective that means "unfastened, flexible, or unconfined." *Lose,* which rhymes with *choose,* is a verb that means to "misplace." Examples:

■ The power failure was linked to a *loose* connection at the switchbox.

■ Because of poor service, the copying machine company may *lose* its contract with our San Francisco office.

MEMO FORMAT

With minor variations, most memoranda (memos) look much the same. The "date/to/from/subject" information hangs at the top left margin, in whatever order your organization requires. These lines help you avoid lengthy introductory information in the memo itself. Figure B–4 outlines the format for a typical memo. Pay special attention to the "subject" line, for it immediately conveys the meaning of the memo to readers. In fact, readers use it to decide when, or if, they will read the complete memo. Be brief, but engage interest.

MODIFIERS

Words, phrases, and even dependent clauses can serve as modifiers in a sentence. Their purpose is to qualify, or add meaning to, other elements in the sentence.

| Facsimile reference |

One or more blank lines

| Date of memo |

| Reader's name (and position, if appropriate) |

| Writer's name (and position, if appropriate) |

| Subject of memo |

One or more blank lines

| Paragraph: Single-spaced (optional—first line indented) |

One blank line

| Paragraph: Single-spaced (optional—first line indented) |

One blank line

| Paragraph: Single-spaced (optional—first line indented) |

One blank line

| Typist's initials (optional—writer's initials before typist's initials) |

One blank line

| Enclosure notation |

One blank line

| Copy notation |

FIGURE B–4
Memorandum format
Source: Technical Writing: A Practical Approach, 4th ed, (p. 241) by W. S. Pfeiffer, 2000, Upper
Saddle River, NJ: Prentice Hall. Reprinted by permission.

Modifiers need to be clearly connected to what they modify. When the connection is not clear, modification errors can occur. Such errors occur most often with phrases that include verbals. There are three types of verbals:

1. **Gerunds:** Nouns that are formed by adding *-ing* to a verb form—for example, "Skiing is his favorite hobby."
2. **Participles:** Adjectives that are formed by adding *-ing* or *-ed* to a verb form—for example, "John is very particular about his writing utensils."
3. **Infinitives:** A phrase that includes the word *to* and a verb root—for example, "To attend college was his main goal."

Modification errors can occur with all verbals. However, they are most common with participles and come in the form of dangling or misplaced modifiers.

■ DANGLING MODIFIERS

When a verbal phrase "dangles," the sentence in which it is used contains no specific word for the phrase to modify. As a result, the meaning of the sentence can be confusing to the reader. You can correct dangling modifiers either by (1) rewording the sentence to give the participle a specific noun to modify (see Revision 1) or by (2) recasting the sentence to remove the participle (see Revision 2).

Original: Using an angle of friction of 20 degrees and a vertical weight of 300 tons, the sliding resistance would be as shown on Table 2.

Revision 1: Using an angle of friction of 20 degrees and a vertical weight of 300 tons, we computed a sliding resistance as shown on Table 2.

Revision 2: If the angle of friction is 20 degrees and the vertical weight is 300 tons, the sliding resistance should be as shown on Table 2.

■ MISPLACED MODIFIERS

When a verbal phrase is misplaced, it appears to refer to a word that it does not modify. At best, misplaced modifiers can lead to unintentionally amusing prose. At worst, they can cause confusion about the agent of action in technical tasks. Correct the error by recasting the sentence, with modifiers clearly connected to what they modify.

Original: Floating peacefully near the oil rig, we saw two humpback whales. [The sentence should indicate the whales are floating.]

Revision: We saw two humpback whales floating peacefully near the oil rig.

Original: Before beginning to dig the observation trenches, we recommend that the contractors submit their proposed excavation program for our review.

Revision: We recommend the following: before the contractors begin digging observation trenches, they should submit their proposed excavation program for our review.

NUMBERS

Like rules for abbreviations, those for numbers may vary from profession to profession and even from company to company. Most technical writing subscribes to the approach that numbers are best expressed in figures (45) rather than words (forty-five). Note that this style may differ from that in other types of writing. Unless the preferences of a particular reader suggest you do otherwise, use the following common rules for numbers:

■ FOLLOW THE 10-OR-OVER RULE

In general, use figures for numbers of 10 or more, words for numbers below 10. Examples:

> Three technicians at the site; 15 reports submitted last month; one rig contracted for the job

■ DO NOT START SENTENCES WITH FIGURES

Begin sentences with the word form of numbers, not with figures. Example:

> Forty-five containers were shipped back to the lab.

■ USE FIGURES AS MODIFIERS

Whether above or below 10, numbers are usually expressed as figures when used as modifiers with units of measurement, time, and money. Examples:

> 4 in., 7 hr, 17 ft-lb, $5 per hour

■ USE FIGURES IN A GROUP OF MIXED NUMBERS

Use only figures when the numbers grouped together in a passage (usually just one sentence) are both above and below 10. Example:

> For that project they assembled 15 samplers, 4 rigs, and 25 large containers.

In other words, this rule argues for consistency within a writing unit.

■ *Use the Figure Form in Illustration Titles*

Use the figure (numeric) form when labeling specific tables and figures in your reports. Example:

Figure 3, Table 14–B

■ *Be Careful with Fractions*

Express fractions as words when they stand alone, but as figures when they are used as a modifier or are joined to whole numbers. Example:

We have completed two-thirds of the project using the 2½-in. pipe.

■ *Use Figures and Words with Numbers in Succession*

When two numbers appear in succession in the same unit, write the first as a word and the second as a figure. Example:

We found fifteen 2-ft pieces of pipe in the bin.

■ *Only Rarely Use Numbers in Parentheses*

Except in legal documents, avoid the practice of placing figures in parentheses after their word equivalents. Example:

The second party will pay the first party forty-five (45) dollars on or before the first of each month.

Note that the parenthetical amount is placed immediately after the figure, not after the unit of measurement.

■ *Use Figures with Dollars*

Use figures with all dollar amounts, with the exception of the context noted in the preceding rule. Avoid cents columns unless exactness to the penny is necessary.

■ *Use Commas in Four-Digit Figures*

To prevent possible misreading, use commas in figures of four digits or more. Example:

15,000; 1,247; 6,003

■ USE WORDS FOR ORDINALS

Usually spell out the ordinal form of numbers. Example:

> The judge informed the first, second, and third choices [not 1st, 2nd, and 3rd choices] in the design competition.

A notable exception is graphics, where space limitations could argue for use of the abbreviated form.

NUMBER OF/TOTAL OF

These two phrases can take singular or plural verbs, depending on the context. Here are two simple rules for correct usage:

1. If the phrase is preceded by *the,* it takes a singular verb because emphasis is placed on the group.
2. If the phrase is preceded by *a,* it takes a plural verb because emphasis is placed on the many individual items.

Examples:

- *The number of* projects going over budget *has* decreased dramatically.
- *The total of* 90 lawyers *believes* the courtroom guidelines should be changed.
- *A number* of projects *have* stayed within budget recently.
- *A total of* 90 lawyers *believe* the courtroom guidelines should be changed.

ORAL/VERBAL

Oral refers to words that are spoken, as in "*oral presentation."* *Verbal* refers to spoken or written language, as opposed to "nonverbal" communication such as body language. Avoid the common error of using *verbal* as a substitute for *oral.* Examples:

- In its international operations, our firm has learned that some countries still rely on *oral* [not *verbal*] contracts.
- Their *oral* [not *verbal*] agreement last month was followed by a *written* [not *verbal*] contract this month.

PARAGRAPH GUIDELINES

Paragraphs have two objectives. First, they must state and then develop a topic. Second, they must maintain reader interest in the main idea. These objectives sometimes seem to work at cross-purposes, for many writers force paragraphs to bear the burden of too much detail. Remember that readers become frustrated by (and even tend to skip over) paragraphs that go beyond six or eight lines.

Here are five rules to help you write clear, thoughtful, and engaging paragraphs:

■ PUT THE MAIN POINT FIRST

Start the paragraph with your main idea. Do not delay or bury the main point in the middle of the paragraph. Readers who skim documents will usually focus on the first sentences of paragraphs. If you fail to put the main point there, they may miss your point entirely.

■ KEEP PARAGRAPHS SHORT

Restrict paragraph length to six or eight lines. Most people do not want to read more than that without a visual break. If your idea requires more thorough treatment, divide the topic at a convenient point and use two paragraphs to develop it.

■ USE LISTS

Take groups of related points and form bulleted or numbered lists within the paragraph. Readers lose patience when information could have been more clearly presented in listings.

■ AVOID EXTENSIVE USE OF NUMBERS

Paragraphs are the worst format for presenting data of any kind, especially numbers that describe costs. Readers may ignore or miss data that are packed into paragraphs. Usually tables or figures would be a clearer and more appropriate format. Also, be aware that some readers may think that cost data couched in paragraph form represent an attempt to hide important information.

■ GIVE READERS SOME VARIETY

Alter the length of paragraphs, to make your writing more visually appealing. Remember that a very short paragraph should emphasize a major point or provide a transition between two longer paragraphs.

PARENTHESES

Use parentheses sparingly because long parenthetical expressions can distract readers. But they are useful for enclosing the following items:

- An abbreviation immediately following the complete term
- A brief explanation that provides extra, subordinate information
- Reference citations within the text

As for punctuation, a period goes after the close parenthesis when the parenthetical information is part of the sentence (as in this sentence). (However, the period goes inside the close parenthesis when the parenthetical information forms its own sentence, as in the sentence you are reading.)

PARTS OF SPEECH

This term refers to the eight main groups of words in English grammar. A word's placement into one of these groups is based upon its function within the sentence, as described here:

NOUNS

Nouns name persons, places, objects, or ideas. The two major categories are (1) proper nouns and (2) common nouns. Proper nouns name specific persons, places, objects, or ideas, and they are capitalized. Examples:

Cleveland, Mississippi River, Apple Computer, Student Government Association, Susan Jones, Existentialism

Common nouns name general groups of persons, places, objects, and ideas, and they are not capitalized. Examples:

trucks, farmers, engineers, assembly lines, philosophy, committee

VERBS

Verbs express action or state of being. As such they give movement to the sentence and form the core of meaning in your writing. Examples:

explore, grasp, write, develop, is, has

PRONOUNS

Pronouns act as substitutes for nouns. Some sample pronoun categories include the following:

- Personal pronouns (I, we, you, she, he)
- Relative pronouns (who, whom, that, which)

- Reflexive and intensive pronouns (myself, yourself, itself)
- Demonstrative pronouns (this, that, these, those)
- Indefinite pronouns (all, any, each, anyone)

ADJECTIVES

Adjectives modify nouns and thus add meaning to them. Examples:

> horizontal, stationary, green, large, simple

ADVERBS

Adverbs modify verbs, adjectives, other adverbs, or whole statements. Examples:

> soon, generally, well, very, too, greatly

PREPOSITIONS

Prepositions show the relationship between nouns or pronouns (called the objects of prepositions) and other elements in a sentence. Forming a prepositional phrase, the preposition and its object can reveal relationships such as location, time, and direction. Examples:

- They went *over the hill.* (location)
- He left *after the meeting.* (time)
- She walked *toward the office.* (direction)

CONJUNCTIONS

Conjunctions are connecting words that link words, phrases, or clauses. They are called coordinate conjunctions when they link like elements, such as two main clauses. They are called subordinate conjunctions when they link two different elements, such as a dependent clause and an independent clause. Examples:

- Coordinate: and, but, for, nor, or, so, yet
- Subordinate: after, although, as if, because, before, even though, if, now that, once, since, so that, though, unless, until, when, whenever, where, whereas, while

INTERJECTIONS

As an expression of emotion, these words can stand alone (Look out!), or they can be inserted into another sentence (*Oh,* now I understand what you meant in your proposal).

PASSED/PAST

Passed is the past tense of the verb *pass,* whereas *past* is an adjective, preposition, or noun that means "previous" or "beyond" or "a time before the present." Examples:

- He *passed* the survey marker on his way to the construction site.
- The *past* president attended last night's meeting. [adjective]
- He worked *past* midnight on the project. [preposition]
- In the distant *past,* the valley was a tribal hunting ground. [noun]

PER

Coming from the Latin, this word should be reserved for business and technical expressions that involve statistics or measurement—such as "per annum" or "per mile." It should *not* be used as a stuffy substitute for "in accordance with." Examples:

- Her *per diem* travel allowance of $90 covered hotels and meals.
- During the oil crisis years ago, gasoline prices increased by over 50 cents *per gallon.*
- *As you requested* [not *per your request*], we have enclosed brochures on our products.

PER CENT/PERCENT/PERCENTAGE

Per cent and *percent* have basically the same usage and occur with exact numbers. The one word *percent* is preferred. Even more common in technical writing, however, is the use of the percentage sign (%) after numbers. The word *percentage* is only used to express general amounts, not exact numbers. Examples:

- After completing a marketing survey, we discovered that 83 *percent* [or 83%] of our current clients have hired us for previous projects.
- A large *percentage* of the defects can be linked to the loss of two experienced inspectors.

PRACTICAL/PRACTICABLE

Though close in meaning, these two words have quite different implications. *Practical* refers to an action that is known to be effective. *Practicable* refers to an action that can be accomplished or put into practice, without regard for its effectiveness or practicality. Examples:

- His *practical* solution to the underemployment problem led to a 30% increase in employment last year.
- The department head presented a *practicable* response, for it already had been put into practice in another branch.

PRINCIPAL/PRINCIPLE

When these two words are misused, the careful reader will notice. After reading the sections that follow, you will understand why "The principal principle maintained by the principals of the bank is the importance of both interest and principal."

■ PRINCIPAL

Principal can be either a noun or adjective and has three basic uses:

1. As a noun meaning "head official" or "person who plays a major role."
2. As a noun meaning "the main portion of a financial account upon which interest is paid."
3. As an adjective meaning "main or primary."

Examples:

- The *principals* of the firm each purchased 1,000 shares of stock.
- We withdrew the interest—but not the *principal*—from our account.
- We believe that the *principal* reason for contamination at the site is the leaky underground storage tank.

■ PRINCIPLE

Principle is a noun that means "basic truth, belief, or theorem." Examples:

- He acted on his *principles* when he reported a fellow employee for overcharging a customer.
- The *principle* of free speech means the government allows people to speak even when it detests what they are saying.
- Avogadro's *principle* states that if two equal volumes of gases are at the same temperature and pressure, they contain the same number of molecules.

PRONOUN AGREEMENT AND REFERENCE

Pronouns—like *this, she, anyone,* and *they*—are words that substitute for nouns. The nouns to which pronouns refer are called "antecedents" of their respective pronouns. Pronouns provide a useful strategy for varying your style and avoiding needless repetition of nouns. However, their use requires vigilance in avoiding errors in agreement and reference. Here are the main rules:

◼ MAKE PRONOUNS AGREE WITH ANTECEDENTS

Check every pronoun to make certain it agrees with its antecedent in number. That is, both noun and pronoun must be singular, or both must be plural. Of special concern are the pronouns *it* and *they.* Examples:

- Norax, Inc., will complete *its* [not *their*] Argentina project next month.
- The committee released *its* [not *their*] recommendations to all departments.

◼ BE CLEAR ABOUT THE ANTECEDENT OF EVERY PRONOUN

Leave no doubt about what noun a pronoun replaces. Any confusion about the antecedent of a pronoun can change the entire meaning of a sentence. To avoid such reference problems, rewrite a sentence or even repeat a noun rather than use a pronoun. Do whatever is necessary to prevent misunderstanding by your reader.

> **Original:** The gas filters for these tanks are so dirty that they should not be used.

> **Revision:** These filters are so dirty that they should not be used.

◼ FOLLOW THIS WITH A NOUN

A common stylistic error is the vague use of *this* as a pronoun, especially as the subject of a sentence. Sometimes the reference is totally confusing; sometimes it may be clear after several readings; and sometimes it is fairly clear but still a sign of a vague writing style. Using *this* as a pronoun often makes readers want to ask, "This what?" Make the subject of your sentence concrete, either by adding a noun after the *this* or by recasting the sentence.

> **Original:** He talked every day about his upcoming trip to Europe. This irritated his colleagues.

> **Revision:** He talked every day about his upcoming trip to Europe. This chatter irritated his colleagues.

> OR

> His constant talk about his upcoming European vacation irritated his colleagues.

QUOTATION MARKS

In technical writing, you can use quotation marks for three main reasons:

1. To draw attention to particular words
2. To indicate passages taken directly from another source
3. To enclose titles of short documents such as reports or book chapters

As for punctuation, periods and commas go inside quotation marks. Semicolons and colons go outside of quotation marks. Examples:

- He discussed "fuzzy logic" in detail in his presentation.
- In its report, the project management firm indicated that "Plant 624 will be completed well before the May 26 deadline."
- In the body of the proposal, the section entitled "Rationale" gave a justification of the high labor costs.

REGRETTABLY/REGRETFULLY

Regrettably means "unfortunately," whereas *regretfully* means "with regret." When you are unsure of which word to use, substitute the above definitions to determine correct usage. Examples:

- *Regrettably,* the team members omitted their resumes from the proposal.
- Hank submitted his resume to the investment firm, but, *regrettably,* he forgot to include a cover letter.
- I *regretfully* climbed on the plane to return home from Hawaii.

RESPECTIVELY

Some good writers may use *respectively* to connect sets of related information. Yet such usage creates extra work for readers by making them reread previous passages. It is best to avoid *respectively* by rewriting the sentence, as shown in the several following options. Examples:

Original: Appendices A, G, H, and R contain the topographical maps for Sites 6, 7, 8, and 10, respectively.

Revision—Option 1: Appendix A contains the topographical map for Site 6; Appendix G contains the map for Site 7; Appendix H contains the map for Site 8; and Appendix R contains the map for Site 10.

Revision—Option 2: Appendix A contains the topographical map for Site 6; Appendix G for Site 7; Appendix H for Site 8; and Appendix R for Site 10.

Revision—Option 3: Topographical maps are contained in the appendices, as shown in the following list:

Appendix	*Site*
A	6
G	7
H	8
R	10

SEMICOLONS

Consider the semicolon to be a modified period. Its most frequent use is in situations where grammar would allow you to use a period but where stylistic preference may be for a less abrupt connector. Example:

> Five engineers left the convention hotel after dinner; only two returned by midnight.

One of the most common punctuation errors is the comma splice. It occurs when a comma instead of a semicolon or period is used in compound sentences connected by words like *however, therefore, thus,* and *then.* When these connectors separate two main clauses, either use a semicolon or start a new sentence. Example:

> We made it to the project site by the agreed-upon time; however, [or, "time. However,"] the rain forced us to stay in our trucks for two hours.

As noted in the "Lists: Punctuation" entry on page 178, there is another instance in which you might use semicolons. Place them after the items in a list when you are treating the list like a sentence and when any one of the items contains internal commas.

SENTENCE STRUCTURE: TERMINOLOGY

All writers have their own approach to sentence style. Yet every one of us has the same tools with which to work: words, phrases, clauses, and different sentence patterns. This entry defines some basic terms. The next one includes specific guidelines for organizing information in sentences. See the Parts of Speech entry for a complete description of terms for words. Only the most important ones—subjects and verbs—are discussed below.

■ SUBJECTS AND VERBS

The most important parts of the sentence are the subject and the verb:

- *Subject:* names the person doing the action or the thing being discussed (*He* completed the study. / The *figure* contains his data).
- *Verb:* conveys action or state of being (She *visited* the site. / He *was* the manager).

■ PHRASES AND CLAUSES

Words can be grouped into two main units, phrases and clauses:

- *Phrase:* lacks either a subject or a verb and thus must always relate to or modify another part of the sentence (She went *to the office./ As project manager,* he had to write the report).
- *Clause:* has both a subject and a verb. It may stand by itself as a *main clause* and thus can be a complete sentence (*He talked to the group*). Or it may rely on another part of the sentence for its meaning and thus is a *dependent clause* (*After he left the site,* he went home).

■ SENTENCE TYPES

There are four main types of sentences:

- *Simple:* contains one main clause (*He completed his work*).
- *Compound:* includes two or more main clauses connected by conjunctions (*Joe completed his work, but Mary stayed at the office to begin another job*).
- *Complex:* includes one main clause and at least one dependent clause (*After he finished the project, Jamal headed for home*).
- *Compound-complex:* contains at least two main clauses and at least one dependent clause (*After they studied the maps, they left the project site, but they were unable to travel much farther that night*).

SENTENCE STRUCTURE: GUIDELINES

How you write and arrange sentences is much a matter of personal writing style. Yet there are a few fundamental guidelines for all good business and technical writing:

■ PLACE THE MAIN POINT NEAR THE BEGINNING

One way to satisfy this criterion for good sentence style is to avoid excessive use of the passive voice (see the Active and Passive Voice entry). Another way is to avoid lengthy introductory phrases or clauses at the beginnings of sentences. The reader usually wants the most important information first.

Original: After discussing the report submitted by the engineering consultant, it was decided by the committee to open a new office.

Revision: The committee decided to open a new office after discussing the consultant's report.

■ FOCUS ON ONE MAIN CLAUSE IN EACH SENTENCE

When you start stringing together too many clauses with *and* or *but,* you dilute the meaning of your text. An occasional compound or compound-complex sentence is acceptable, just for variety, but most sentences should be simple or complex.

■ VARY SENTENCE LENGTH, BUT AVERAGE NO MORE THAN 15–20 WORDS

Place important points in short, emphatic sentences. Reserve longer sentences for qualifications and support of your main points. If you find your sentences run too long, there is a simple technique for shortening them. First find some long compound sentences connected by conjunctions like *and, but,* or *so.* Then separate them into shorter sentences by breaking them at the conjunction. Another strategy is to break sentences into two by separating the main point being made from supporting information.

Original: Last week the human resources staff announced that the company is offering three new health plans because employees have complained about the cost of the previous plans and the poor service provided.

Revision: Last week the human resources staff announced that the company is offering three new health plans. The change came about because of the cost and poor service of previous plans.

SET/SIT

Like *lie* and *lay,* these two verbs are distinguished by form and use. Here are the basic differences:

1. *Set* means "to place in a particular spot" or "to adjust." It is a transitive verb and thus takes a direct object to which it conveys action. Its main

parts are *set* (present), *set* (past tense), *set* (past participle), and *setting* (present participle).

2. *Sit* means "to be seated." It is usually an intransitive verb and thus does not take a direct object. Its main parts are *sit* (present), *sat* (past), *sat* (past participle), and *sitting* (present participle). It can be transitive when used casually as a direction to be seated. ("Sit yourself down and take a break.")

Examples:

- He *set* the computer on the table yesterday.
- While *setting* the computer on the table, he sprained his back.
- The technician had *set* the thermostat at 75 degrees.
- She plans to *sit* exactly where she sat last year.
- While *sitting* at her desk, she saw the computer.

SEXIST LANGUAGE: AVOIDING IT

Language usually *follows* changes in culture, rather than *anticipating* such changes. A case in point is today's shift away from sexist language in business and technical writing—indeed, in all writing and speaking. The change reflects the increasing number of women entering previously male-dominated professions such as engineering, management, medicine, and law.

This section defines sexist and non-sexist language and suggests ways to avoid using gender-offensive language in your writing.

■ *BACKGROUND ON SEXISM AND LANGUAGE*

Sexist language is the use of wording, especially pronouns like *he* or *him*, to represent positions or individuals who could be either men or women. For many years, it was common to use *he, his, him,* or other male words in sentences such as those that follow.

Examples of poor usage:

- The operations specialist should check page 5 of his manual before flipping the switch.
- Every physician was asked to renew his membership in the medical association before next month.
- Each new student at the military academy was asked to leave his personal possessions in the front hallway of the administrative building.

These examples of obsolete usage assume that male pronouns and adjectives can stand for any persons—male or female. There are two good reasons to avoid such usage:

1. The entry of many more women into the professions has called attention to the exclusionary nature of using male pronouns for indefinite use. Many readers will be bothered by such writing.

2. The use of male pronouns in an indefinite context can be viewed as one way women are constrained from achieving equal status in the professions and, generally, in the culture. That is, such use of male pronouns fosters sexism in the society as a whole.

If you fail to rid your writing of sexist language, you risk drawing more attention to your style than to your ideas. Common sense argues for following some basic style rules to avoid sexist language.

■ GUIDELINES FOR NON-SEXIST LANGUAGE

This section offers techniques to help you shift from sexist to non-sexist language. Some of these strategies may not suit your taste in writing style; choose the ones that work for you.

Avoid Personal Pronouns Altogether

An easy way to avoid problems with sexist language is to delete unnecessary pronouns from your writing.

> **Sexist Language:** During his first day on the job, any new employee in the toxic-waste laboratory must report to the company doctor for his employment physical.

> **Non-Sexist Language:** During the first day on the job, each new employee in the toxic-waste laboratory must report to the company doctor for a physical.

Use Plural Pronouns Instead of Singular

Often you can shift from singular to plural pronouns without changing meaning.

> **Sexist Language:** Each geologist should submit his time sheet by noon on the Thursday before checks are issued.

> **Non-Sexist Language:** All geologists should submit their time sheets on the Thursday before checks are issued.

Note also that you may encounter sexist language that inappropriately uses female pronouns in a generic manner. Such usage is just as flawed as male-oriented language.

> **Sexist Language:** Each nurse should make every effort to complete her rounds each hour.

Non-Sexist Language: Nurses should make every effort to complete their rounds each hour.

Alternate Masculine and Feminine Pronouns

You can avoid sexist use by alternating *he* and *him* with *she* and *her*. However, do not switch pronoun use within too brief a passage, such as a paragraph or page. Instead, change every few pages, or every section or chapter. The drawback of this technique is that the alternating use of masculine and feminine pronouns tends to draw attention to itself. Also, the writer must work hard to balance the use of masculine and feminine pronouns, in a sense to give "equal treatment."

Use Forms Like *He or She*

This technique requires the writer to include both genders of pronouns each time indefinite use is required.

Sexist Language: The president made it clear that each branch manager will be responsible for the balance sheet of his respective office.

Non-Sexist Language: The president made it clear that each branch manager will be responsible for the balance sheet of his or her respective office.

However, it is important to note that this style is bothersome to many writers who feel that *his or her* is wordy and awkward. Many readers are bothered even more by the forms *he/she, his/her,* and *him/her.*

Shift to Second-Person Pronouns

Change to *you* and *your,* which are words without any sexual bias. For this technique to be effective, you need to be writing documents in which it is appropriate to use an instructions-related "command" tone associated with the use of *you.*

Sexist Language: After selecting her insurance option in the benefit plan, each new nurse should submit her paperwork to the Human Resources Department.

Non-Sexist Language: Submit your paperwork to the Human Resources Department after selecting your insurance option in the benefit plan.

Be Especially Careful of Titles and Letter Salutations

When you do not know how a particular woman prefers to be addressed, always use *Ms.* Even better, call the person's employer to ask whether *Miss, Mrs., Ms.,* or some other title is appropriate. (When calling, also check on the correct spelling of the person's name and her current job title.) When you do not know who will read your letter, never use *Dear Sir* or *Gentlemen* as a generic greeting. *Dear Sir or Madam* is also inappropriate. Instead, call the organization and get the name of a particular person to whom you can direct your letter. If you must write to a group of people, replace the generic greeting with an "attention" line that denotes the name of the group.

Sexist Language: Dear Sir: [to a collective audience]

Non-Sexist Language: Attention: Admissions Committee

Sexist Language: Dear Miss Finnigan: [to a single woman for whom you can determine no title preference]

Non-Sexist Language: Dear Ms. Finnigan:

SIC

Latin for *thus,* this word is most often used when a quoted passage contains an error or other point that might be questioned by the reader. Inserted within brackets, *sic* indicates that the word or phrase before the *sic* was included in the original passage—and that it was not introduced by you. In other words, *sic* should be included when it is important that the reader be presented with a technically accurate quotation. Examples:

- The customer's letter to our Sales Department claimed that "there are too [*sic*] or three main flaws in the product."
- The proposal we received includes the following statement: "Work on each project will be overseen by a principle [*sic*] of the firm."

SPELLING

All writers have words they find difficult to spell, and some writers have major spelling problems. Spell-checking software can help solve the problem. Yet you still must remain vigilant during the proofreading stage. Misspelled words in an otherwise well-written document cause readers to question the quality of the document and the professionalism of the writer.

This entry includes a list of commonly misspelled words. However, you should keep your own list of words you often have trouble spelling. Most of us have a relatively short list of words that give us repeated difficulty.

absence	accommodate
accessible	accumulate
accustomed	dilemma

achievement

acknowledgment

acquaintance

admittance

advisable

aisle

allotting

analysis

analyze

Arctic

athlete

athletic

awful

basically

believable

benefited

bulletin

calendar

career

changeable

channel

column

commitment

committee

compatible

compelled

conscience

conscientious

conscious

controlled

convenient

definitely

dependable

descend

disappear

disappoint

disaster

disastrous

efficient

eligible

embarrass

endurance

environment

equipment

equipped

essential

exaggerate

existence

experience

familiar

favorite

February

foreign

foresee

forfeit

forty

fourth

genius

government

guarantee

guidance

handicapped

harass

height

illogical

incidentally

independence

indispensable

ingenious	occurrence
initially	omission
initiative	pamphlet
insistence	parallel
interfered	pastime
interference	peculiar
interrupt	possess
irrelevant	practically
judgment	preference
knowledge	preferred
later	privilege
latter	profession
liable	professor
liaison	pronunciation
library	publicly
lightning	quantity
likely	questionnaire
loneliness	recession
maintenance	reference
manageable	safety
maneuver	similar
mathematics	sincerely
medieval	specifically
mileage	subtle
miscellaneous	temperament
misspelled	temperature
mortgage	thorough
movable	tolerance
necessary	transferred
noticeable	truly
nuisance	undoubtedly
numerous	unmistakably
occasionally	until
occurred	useful

usually	wholly
valuable	writing
various	written
vehicle	

STATIONARY/STATIONERY

Stationary means "fixed" or "unchanging," whereas *stationery* refers to paper and envelopes used in writing or typing letters. Examples:

- To perform the test correctly, one of the workers had to remain *stationary* while the other one moved around the job site.
- When she began her own business, Julie purchased *stationery* with her new logo on each envelope and piece of paper.

SUBJECT-VERB AGREEMENT

Subject-verb agreement errors are common in technical writing. They occur when writers fail to make the subject of a clause agree in number with the verb. ("The company president, along with his wife, are attending the meeting.") The verb *are* is *not* correct. The subject is *president,* so the verb should be *is.*

If you have agreement problems in your own writing, you can solve them by (1) isolating the subjects and verbs of all clauses in the document and (2) making certain that they agree. The main subject-verb agreement rules follow:

■ SUBJECTS CONNECTED BY AND TAKE PLURAL VERBS

This rule applies to two or more words or phrases that, together, form one subject phrase. Example:

> The site preparation section and the foundation design portion of the report are to be written by the same person.

■ VERBS AFTER EITHER/OR AGREE WITH THE NEAREST SUBJECT

Subjects connected by *either/or* (or *neither/nor*) confuse many writers, but the rule is very clear. Your verb choice depends on the subject nearest the verb.

Examples:

- He told his group that either the three reports or the proposal was to be completed this week.
- He told his group that either the proposal or the three reports were to be completed this week.

In the first example, the part of the subject closest to the verb is *proposal;* thus the correct verb form is *was*. In the second example, the part of the subject closest to the verb is *reports;* thus the correct verb form is *were*. Though both sentences are grammatically correct, the second is preferred because readers often view mixed *either/or* constructions as plurals. Singular verbs may sound awkward in such sentences.

■ VERBS AGREE WITH THE SUBJECT, NOT WITH THE SUBJECTIVE COMPLEMENT

Sometimes called a predicate noun or adjective, a subjective complement renames the subject and occurs after verbs like *is, are, was,* and *were*. Examples:

- The topic of the report is our high profits last year.
- Our high profits last year are the topic of the report.

■ PREPOSITIONAL PHRASES DO NOT AFFECT MATTERS OF AGREEMENT

As long as, in addition to, and *as well as* are prepositions, not conjunctions. A verb agrees with its subject, not with the object of a prepositional phrase. Example:

> The manager of human resources, as well as the technical managers, is supposed to meet with the three applicants.

■ COLLECTIVE NOUNS USUALLY TAKE SINGULAR VERBS

Collective nouns have singular form but usually refer to a group of persons or things (for example, *team, committee,* or *crew*). When a collective noun refers to a group as a whole, use a singular verb. Example:

> The project crew was ready to complete the assignment.

Occasionally, a collective noun refers to the members of the group acting in their separate capacities. In this case, either use a plural verb or, to avoid awkwardness, reword the sentence. Examples:

- The crew were not in agreement about the survey locations. OR
- Members of the crew were not in agreement about the survey locations.

■ *Foreign Plurals Usually Take Plural Verbs*

Although usage is gradually changing, most careful writers still use plural verbs with *data, strata, phenomena, media,* and other irregular plurals. Example:

> The data he requested are incorporated into the three tables.

■ *Indefinite Pronouns Like Each and Anyone Take Singular Verbs*

Writers often fail to follow this rule when they make the verb agree with the object of a prepositional phrase, instead of with the subject. Examples:

- Each of the committee members are ready to adjourn. (incorrect)
- Each of the committee members is ready to adjourn. (correct)

THERE/THEIR/THEY'RE

These words can be easily misused by accident, even by those who know their correct use. Do a careful job of editing to make sure you have followed these guidelines:

1. Use *there* as an adverb that indicates direction or that is part of the phrases *there is* and *there are.*
2. Use *their* to indicate possession or ownership.
3. Use *they're* as a contraction for *they are.*

Examples:

- You'll find the new forms over *there* in the file cabinet.
- *There is* only one road to that village in Nepal.
- The secretaries and word processors were proud of *their* successful effort to rewrite the company's style manual.
- As for the board members, *they're* all planning to attend the reception tonight.

TO/TOO/TWO

To is either part of the infinitive verb form ("to go") or a preposition in a prepositional phrase ("to the office"). *Too* is an adverb that either suggests an excessive amount or that conveys the meaning "also." *Two* is a noun or adjective that stands for the numeral 2. Examples:

- He volunteered *to* go [infinitive verb] *to* Alaska [prepositional phrase] *to* work [another infinitive verb form] on the project.
- Stephanie explained that the proposed hazardous waste dump would pose *too* many risks to the water supply. Scott made this point *too*.
- When *two* water mains broke, the city experienced a crisis.

UTILIZE/USE

Utilize is simply a long form for the preferred verb *use*. Although some verbs that end in *-ize* are useful words, most are simply wordy substitutes for shorter forms. As some writing teachers say, "Why use *utilize* when you can use *use*." Examples:

- She *used* [not *utilized*] the Internet for her research.
- Three different drills were *used* [not *utilized*] to drill through the rock.

WHICH/THAT

Which is used to introduce nonrestrictive clauses, which are defined as clauses not essential to meaning (as in this sentence). Note that such clauses require a comma before the *which* and a slight pause in speech. *That* is used to introduce restrictive clauses that are essential to the meaning of the sentence (as in this sentence). Note that such clauses have no comma before the *that* and are read without a pause. *Which* and *that* can produce different meanings, as the following examples show:

- Our benefits package, *which* is the best in our industry, includes several options for medical care.
- The benefits package *that* our firm provides includes several options for medical care.
- My daughter's school, *which* is in Cobb County, has an excellent math program.
- The school *that* my daughter attends is in Cobb County and has an excellent math program.

Note that the above examples with *that* might be considered wordy by some readers. Indeed, such sentences often can be made more concise by deleting the *that* introducing the restrictive clause. However, delete *that* only if you can do so without creating an awkward, choppy sentence.

WHO/WHOM

Who is a subjective form that can only be used in the subject slot of a clause; *whom* is an objective form that can only be used as a direct object or other non-subject noun form of a sentence. Examples:

- The man *who* you said called me yesterday is a good customer of the firm. [The clause "who . . . called me yesterday" modifies *man*. Within this clause, *who* is the subject of the verb *called*. Note that the subject role of *who* is not affected by the two words "you said," which interrupt the clause.]
- They could not remember the name of the person *whom* they interviewed. [The clause "whom they interviewed" modifies "person." Within this clause, *whom* is the direct object of the verb *interviewed*. Grammatically, the clause could be turned around to read "they interviewed whom."]

WHO'S/WHOSE

Who's is a contraction that replaces *who is*, whereas *whose* is a possessive adjective. Examples:

- *Who's* planning to attend the annual meeting?
- Susan is the manager *who's* responsible for training.
- *Whose* budget includes training?
- Susan is the manager *whose* budget includes training.

WORDINESS

Some experts believe that careful attention to conciseness would shorten technical documents considerably—perhaps as much as 40% to 50%. As a result, reports and proposals would take less time to read and cost less to produce. This entry offers several techniques for reducing verbiage without changing meaning.

REPLACE ABSTRACT NOUNS WITH VERBS

Concise writing depends more on verbs than it does on nouns. Sentences that contain abstract nouns, especially ones with more than two syllables, can be shortened by focusing on strong verbs. By converting an abstract noun to an action verb, you can eliminate extra words in each wordy sentence below:

Original: The *acquisition* of the property was accomplished through long and hard negotiations.

Revision: The property was *acquired* through long and hard negotiations.

Original: *Confirmation* of the contract occurred yesterday.

Revision: The contract was *confirmed* yesterday.

Original: *Exploration* of the region had to be effected before the end of the year.

Revision: The region had to be *explored* before the end of the year.

Original: *Replacement* of the transmission was achieved only three hours before the race.

Revision: The transmission was *replaced* only three hours before the race.

You can see that abstract nouns often end with *-tion* or *-ment*. Although such words are not always "bad" usage, they do cause problems when they replace the action verbs from which they are derived. In the following examples, abstract nouns are listed in the left column and the preferred verb forms are listed in the right column:

assessment	assess
classification	classify
computation	compute
delegation	delegate
development	develop
disbursement	disburse
documentation	document
elimination	eliminate
establishment	establish
negotiation	negotiate
observation	observe
requirement	require
verification	verify

■ *SHORTEN WORDY PHRASES*

Many wordy phrases have worked their way into business and technical writing. Such weighty expressions add extra words and rob prose of clarity. Here are some of the culprits, along with their concise substitutes:

afford an opportunity to	permit
along the lines of	like
an additional	another
at a later date	later
at this point in time	now
by means of	by
come to an end	end
due to the fact that	because
during the course of	during
for the purpose of	for
give consideration to	consider
in advance of	before
in the amount of	for
in the event that	if
in the final analysis	finally
in the proximity of	near
prior to	before
subsequent to	later
with regard to	about

■ *REPLACE LONG WORDS WITH SHORT ONES*

In school, most students rightly are taught to experiment with more sophisticated words. While this effort helps build vocabularies, it also can encourage the life-long tendency to use long words when short ones will do. As a rule, the following long words in the left column should be replaced by the short words in the right column:

advantageous	helpful
alleviate	lessen, lighten
approximately	about
cognizant	aware

commence	start, begin
demonstrate	show
discontinue	end, stop
endeavor	try
finalize	end, complete
implement	do, carry out
initiate	start, begin
inquire	ask
modification	change
prioritize	rank, rate
procure	buy
terminate	end, fire
transport	move
undertake	try, attempt
utilize	use

■ *LEAVE OUT CLICHÉS*

Clichés are worn-out expressions that add extra words to your writing. Though they once were fresh phrases, they became clichés when they no longer conveyed their original meaning. You can make writing more concise by replacing them with a good adjective or two. Here are some clichés to avoid:

as plain as day
ballpark figure
efficient and effective
few and far between
last but not least
leaps and bounds
needless to say
reinvent the wheel
skyrocketing costs
step in the right direction

■ *WRITE LIKE YOU TALK*

Much wordiness results from talking around the topic. Sometimes called *circumlocution,* this flaw arises from a tendency to write indirectly. It can be avoided by reading passages aloud. Almost invariably, the sound of the words will make problems of wordiness quite apparent. Ask yourself, "How would I say this if I were talking to the reader in person?" Asking this question helps condense all kinds of inflated language. However, remember that even direct writing must retain a tactful, diplomatic tone when it contains negative or sensitive information.

Original: We would like to suggest that you consider directing your attention toward completing the project before the commencement of the seasonal monsoon rains in the region of the project area.

Revision: We suggest you complete the project before the monsoons begin.

Original: At the close of the last phase of the project, a bill for your services should be expedited to our central office for payment.

Revision: After the project ends, please send your bill immediately to our central office.

Original: It is possible that the well-water samples collected during our investigation of the well on the site of the subdivision could possibly contain some chemicals in concentrations higher than is allowable according to the state laws now in effect.

Revision: Our samples from the subdivision's well might contain chemical concentrations beyond that permitted by the state.

■ *AVOID* THERE ARE, IT IS, *AND SIMILAR CONSTRUCTIONS*

There are and *it is* are not good substitutes for concrete subjects and action verbs. Such constructions delay the delivery of information about who or what is doing something. They tend to make your writing lifeless and abstract. Avoid them by creating (1) main subjects that are concrete nouns and (2) main verbs that are action words. The revised passages below give readers a clear idea of "who is doing what" in the subject and verb positions.

Original: There are many McDuff projects that could be considered for design awards.

Revision: Many McDuff projects could be considered for design awards.

Original: It is clear to the hiring committee that writing skills are an important criterion for every technical position.

Revision: The hiring committee believes that writing skills are an important criterion for every technical position.

Original: There were seven people who attended the meeting at the client's office in Charlotte.

Revision: Seven people attended the meeting at the client's office in Charlotte.

◼ CUT OUT EXTRA WORDS

Delete extra words or redundant phrasing in your work. Sometimes the problem comes in the form of needless connecting words, like *to be* or *that*. Other times it shows itself as points made earlier that do not need repeating. Delete extra words when their use (1) does not add a necessary transition or (2) does not provide new information. (One important exception is the intentional repetition of main points for emphasis, as in repeating important conclusions in different parts of a report.) The examples below display a variety of examples of wordy writing, with corrections made in longhand:

- Preparing the client's final bill involves ~~the~~ checking ~~of~~ all ^project^ invoices ⊙ ~~for the project.~~
- The report examined what the ABC project manager considered ~~to be~~ a technically acceptable risk.
- During ~~the course of~~ its field work, the ABC team will ~~be engaged in the process of~~ reviewing all ~~of the~~ notes ~~that have been~~ accumulated in previous studies.
- ~~Because of his position~~ as head of ~~the~~ ABC public relations group, ~~at ABC,~~ he planned ~~such that he would be able~~ to attend the meeting.
- She believed ~~that the~~ recruiting ~~of~~ more minorities for the technical staff is essential.
- The department must determine its ~~aims and~~ goals so that they can be included ^ABC's 2003^ in ~~the~~ annual strategic plan. ~~produced by ABC for the year of 2002.~~
- Most ABC managers ~~generally~~ agree that all ~~of the~~ company's employees ~~at all of the offices~~ deserve ~~at least~~ some ~~degree of~~ training each year ⊙ ~~that they work for the firm.~~

WORDS THAT AREN'T

Many non-words have worked their way into informal and formal usage. Listed below are some common examples.

Don't Use	Use
alot	a lot
anyways	anyway
anywheres	anywhere
can not	cannot
complected	complexioned
could of	could have

different than	different from
enthused	enthusiastic
hisself	himself
in regards to	regarding *or* about
irregardless	regardless
refer back to	refer to
repeat again	repeat
somewheres	somewhere
theirselves	themselves
towards	toward
use to *or* suppose to	used to *or* supposed to
would of	would have

YOUR/YOU'RE

Your is an adjective that shows ownership, whereas *you're* is a contraction for *you are*. Examples:

- *Your* office will be remodeled next week.
- *You're* responsible for giving performance appraisals.

Index